Getting on with maths 2

Contents of Book 1

GETTING ON WITH
MATHS

Derek Farmer B.Tech.
Head of Mathematics
Swinton Comprehensive School
Rotherham

Longman

To Rachel, Matthew and Sam

LONGMAN GROUP LIMITED
Longman House, Burnt Mill, Harlow, Essex CM20 2JE, England
and Associated Companies throughout the world

First published 1985
ISBN 0 582 22436 5

Photoset in 10/12pt Univers Medium

Printed in Great Britain
by Butler & Tanner Limited, Frome and London

Contents

Preface

Many text books during recent years have been of the intuitive kind, which relies on the students' reading skills and initiative. I do not favour this approach, but fully believe that mathematics teaching should be teacher-centred and that it is the class teacher's responsibility to generate interest and stimulate enthusiasm in the subject.

I have produced two text books that provide a large store of useful material covering all the main subject matter for the present CSE examinations and the parallel levels of the 16+ examination.

Each chapter is subdivided into sections containing a particular aspect of the work. Where possible, chapters follow a logical sequence, such as 'The laws of indices', 'Standard index notation', then 'Logarithms'. Most sections have a number of worked examples showing how a variety of different question types may be answered. These are useful as a guide to the student, but methods used are by no means exhaustive. Some of the solutions may seem long, but I feel that it is necessary to show a large number of small but logical stages in reaching the final answer, since it is very easy to lose the understanding of the reader by making large unexplained steps. A class teacher can easily back-track on a solution, but not so a text book.

The exercises in each chapter are designed to provide plenty of mechanical practice and there is a comprehensive range of problem-style questions, the most difficult of which should stretch the more able pupil studying mathematics at this level.

In conclusion, I have also written two Answer Books, which reproduce all the diagrams and graphs that are required for any specific question. I feel that this is an important time-saving aid for the class teacher, who is usually requested to repeat the same diagram several times over.

D. Farmer 1985

1 Symmetry

Reflectional symmetry

A figure possesses reflectional symmetry if a straight line can be drawn through it so that the shape on one side of the line is a mirror image of the shape on the opposite side of the line.

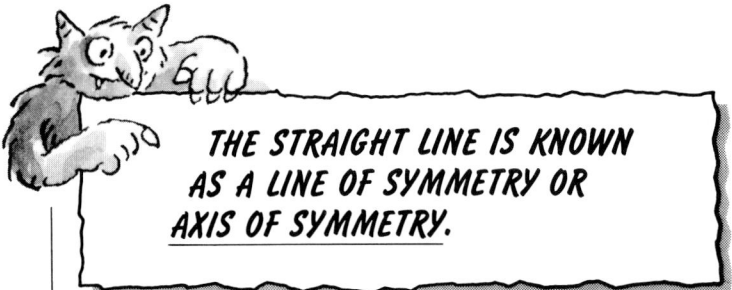

THE STRAIGHT LINE IS KNOWN AS A LINE OF SYMMETRY OR AXIS OF SYMMETRY.

Some figures have more than one line of symmetry.

Example 1

Figure 1.1 shows a shape with one line of symmetry.

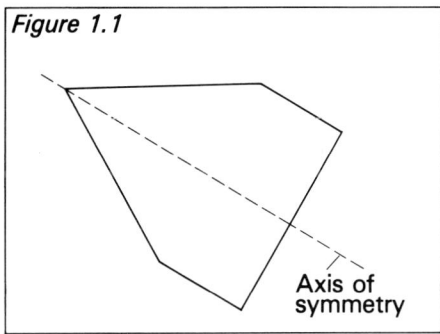

Figure 1.1

Axis of symmetry

Example 2

Figure 1.2 shows a shape with two lines of symmetry.

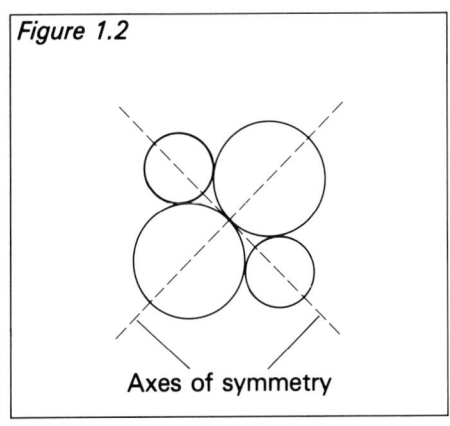

Figure 1.2

Axes of symmetry

Exercise 1.1

Draw in any line(s) of symmetry which may exist in the figures shown in questions 1 to 15. Complete the shapes in questions 16 to 20, where the line of symmetry is given.

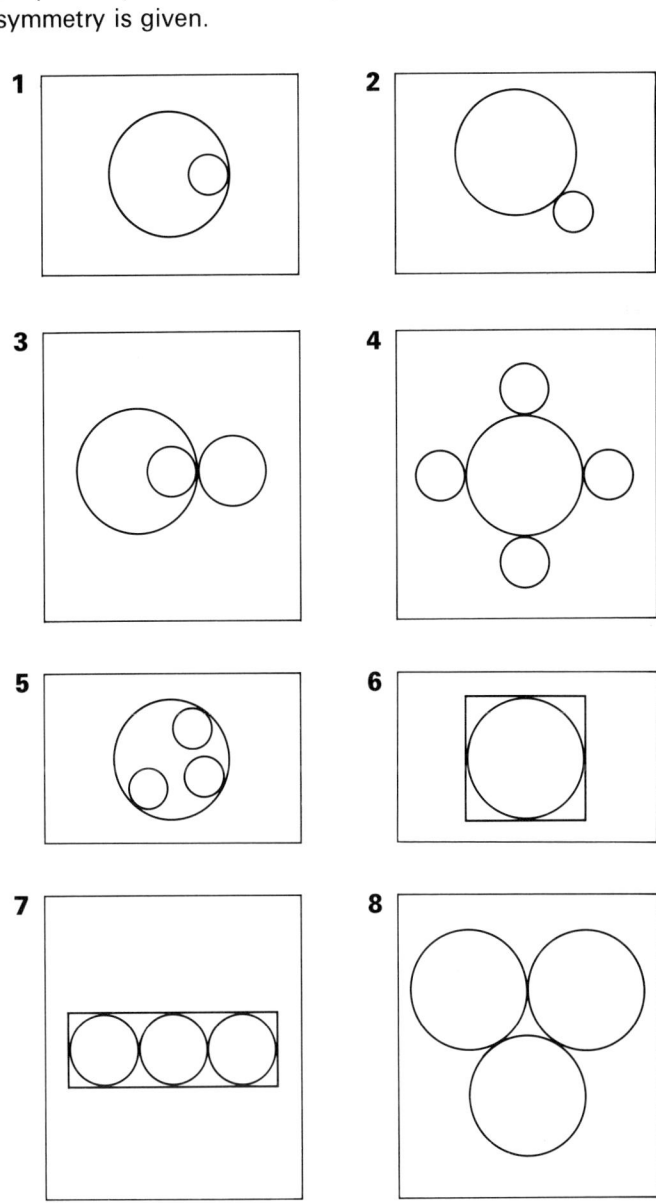

11

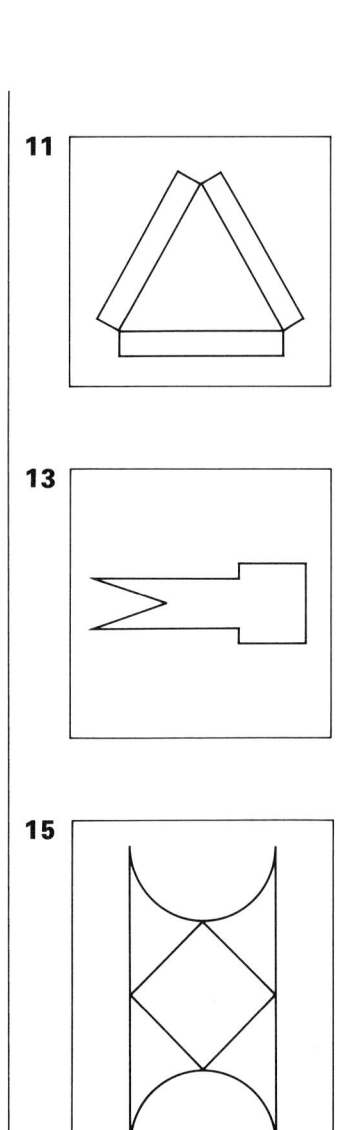

12

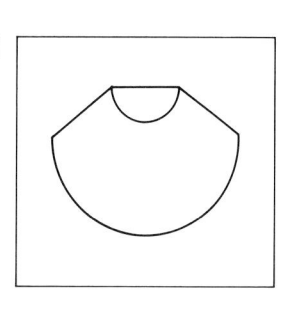

13

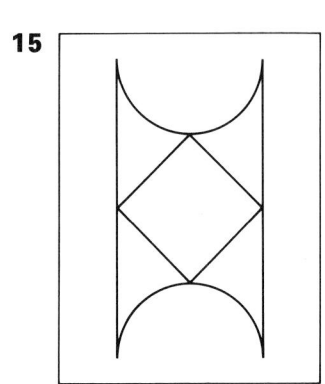

14

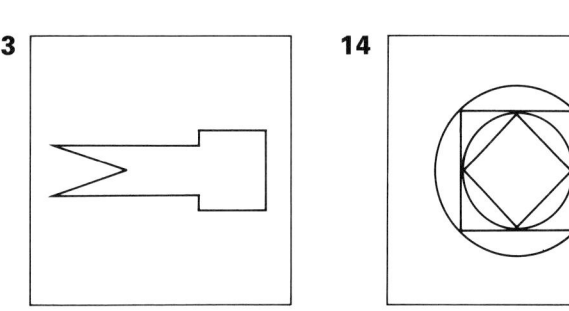

15

16

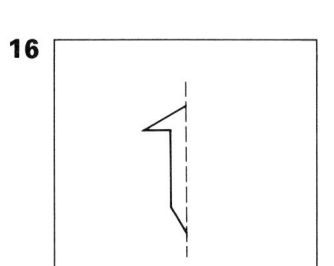

17

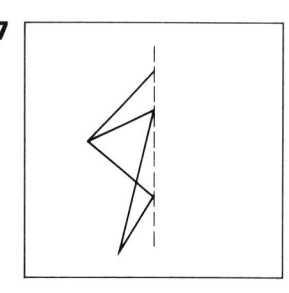

18

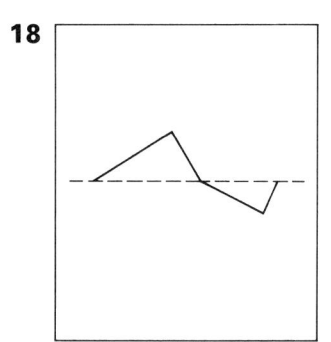

19

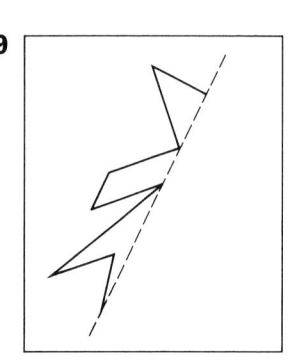

20

Rotational symmetry

A shape possesses rotational symmetry if it can be turned through an angle, about some centre of rotation, so that the initial and final positions of the outline are identical.

> THE NUMBER OF TIMES THAT A SHAPE CAN RETAIN ITS ORIGINAL OUTLINE IN ANY ONE COMPLETE TURN IS CALLED ITS _ORDER_ OF ROTATIONAL SYMMETRY.

Example 3

Draw in the centre of rotation for a square and state its order of symmetry.

Solution

See Figure 1.3.

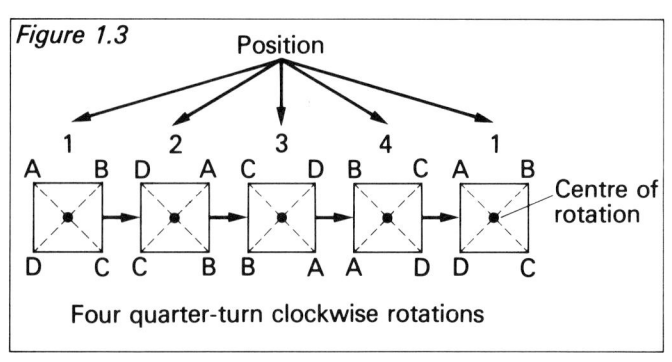

Figure 1.3

Four quarter-turn clockwise rotations

The square has an order of rotational symmetry of 4.

Half-rotational symmetry

A shape possesses half-rotational symmetry or point symmetry if a centre of rotation can be chosen such that when the shape has been turned through 180°, the initial and final positions of the outline are identical.

Examples 4

The large dots show the centres of half-rotational symmetry of the shapes in Figures 1.4 and 1.5.

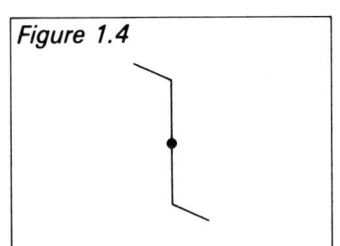

Figure 1.4

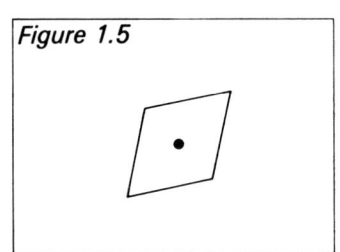

Figure 1.5

ANY SHAPE WHICH HAS TWO AXES OF SYMMETRY AUTOMATICALLY POSSESSES *HALF-ROTATIONAL SYMMETRY.*

Exercise 1.2

In questions 1 to 15 mark on any centres of rotation with a large dot. State whether the shape possesses nil, half-rotational or rotational symmetry (giving the order).

In questions 16 to 20 complete the shapes when the centres of half-rotational symmetry are given.

1

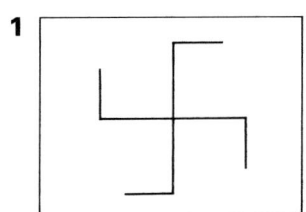

2

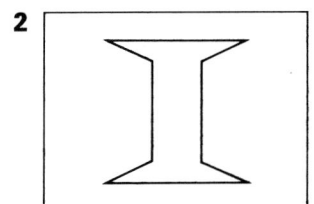

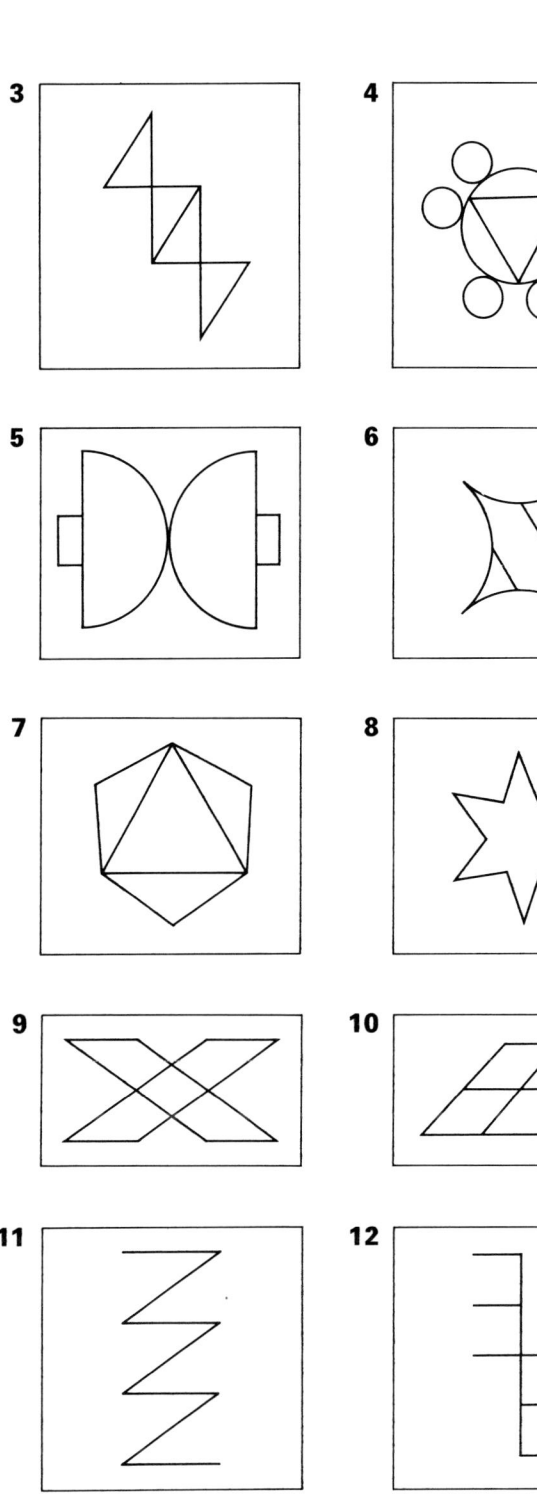

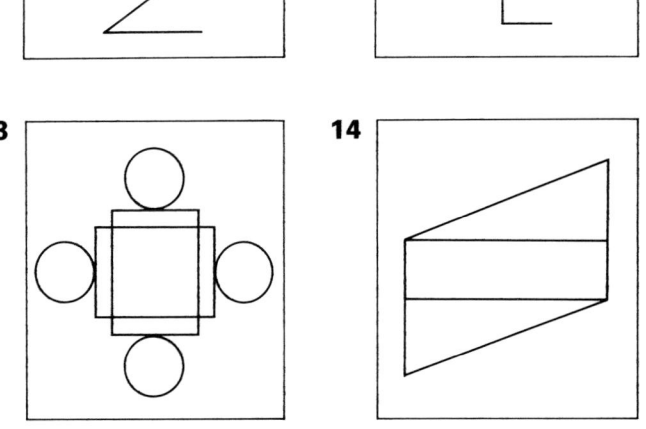

15

16

17

18

19

20

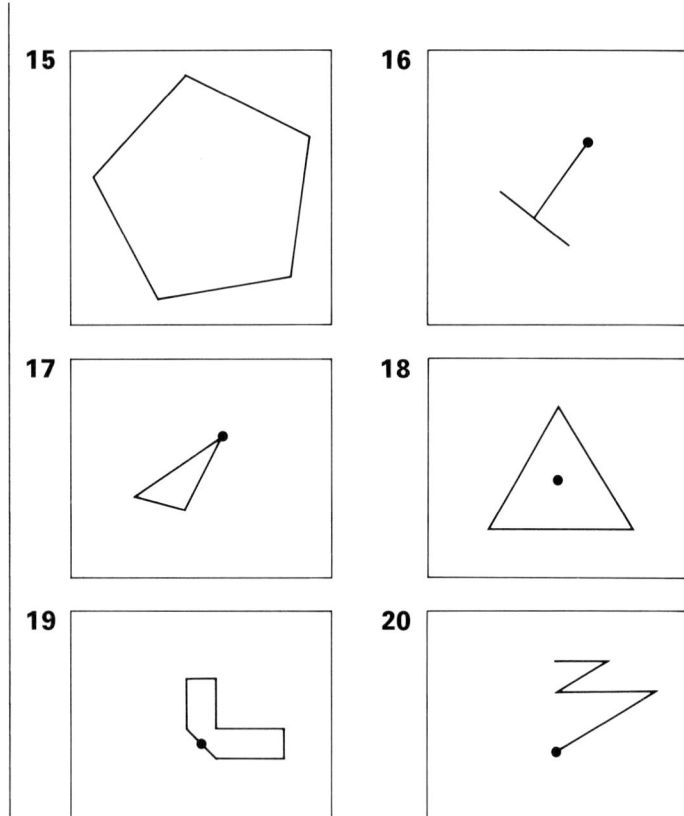

2 Transformations

Translations

Figure 2.1 shows a letter E with reference to the cartesian x–y axes.

> A TRANSLATION CAN BE OBTAINED BY
> (a) MOVING LETTER E PARALLEL TO THE x-AXIS (E_1),
> (b) MOVING LETTER E PARALLEL TO THE y-AXIS (E_2).
> (c) MOVING LETTER E PARALLEL TO BOTH THE x AND y AXES (E_3).

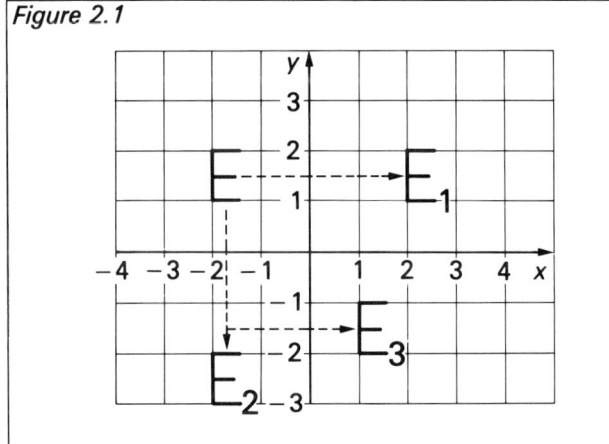

Figure 2.1

Example 1

Copy triangle A in Figure 2.2 and then redraw it after a translation of $\begin{pmatrix} 2 \\ -5 \end{pmatrix}$ has been applied.

Solution

The translation $\begin{pmatrix} 2 \\ -5 \end{pmatrix}$ means move the triangle 2 units parallel to the x-axis in a *positive* direction (dotted

triangle) followed by 5 units parallel to the y-axis in a *negative* direction. This results in the final solution, triangle B.

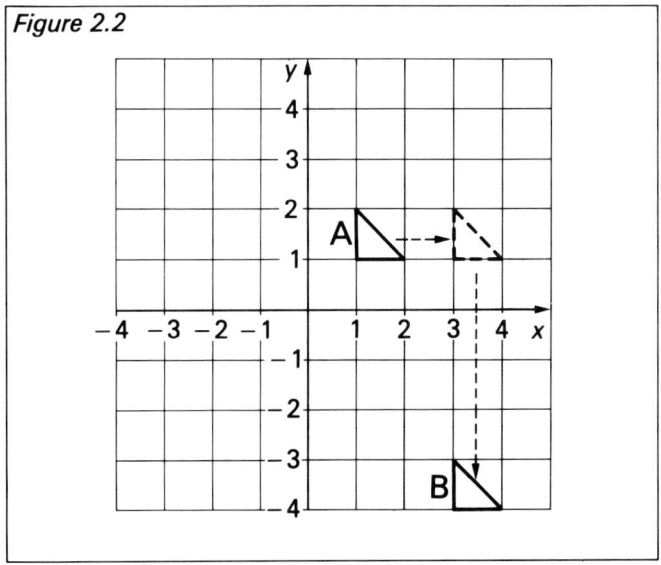

Figure 2.2

Exercise 2.1

1 Copy and complete Figure 2.3 when the triangle
   (a) undergoes a translation of 4 units parallel to the x-axis
   (b) undergoes a translation of 2 units parallel to the y-axis
   (c) undergoes a translation of 3 units parallel to the x-axis followed by -2 units parallel to the y-axis.

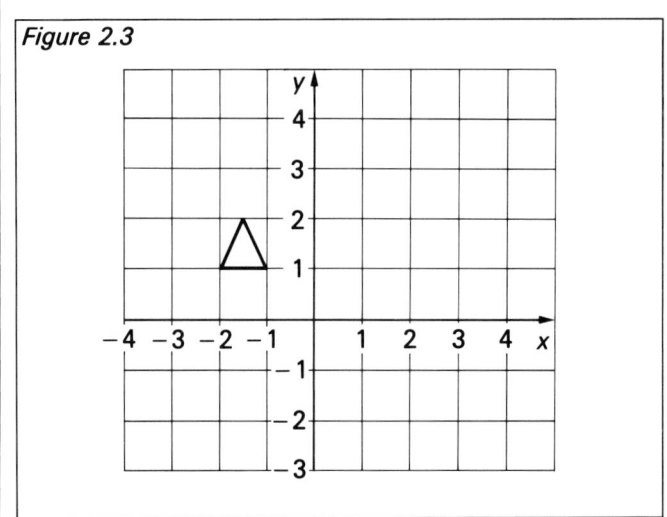

Figure 2.3

2 Copy and complete Figure 2.4.

Figure 2.4

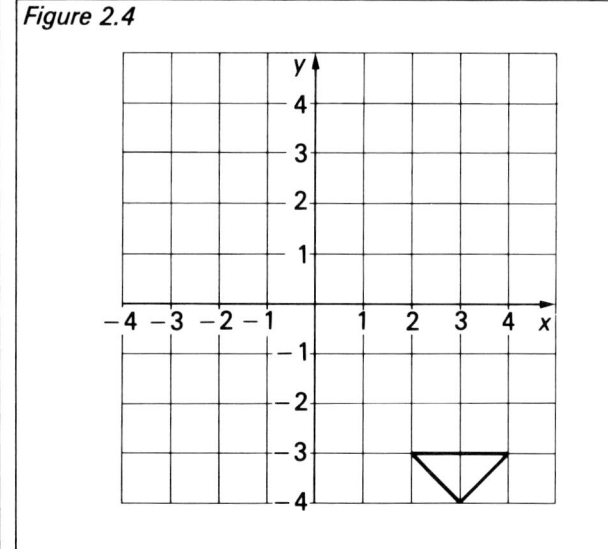

Draw in the following translations when they are applied to the triangle
(a) −6 units parallel to the *x*-axis
(b) 7 units parallel to the *y*-axis
(c) −4 units parallel to the *x*-axis followed by 5 units parallel to the *y*-axis.

3 Draw cartesian axes and plot point M(4, 2).

(a) Plot M′ after applying translation $\begin{pmatrix} -5 \\ 0 \end{pmatrix}$ to point M.

(b) Plot M″ after applying translation $\begin{pmatrix} 0 \\ -3 \end{pmatrix}$ to point M′.

(c) Plot M‴ after applying translation $\begin{pmatrix} 2 \\ -1 \end{pmatrix}$ to point M″.

4 What translation has taken K to K′ in Figure 2.5?

Figure 2.5

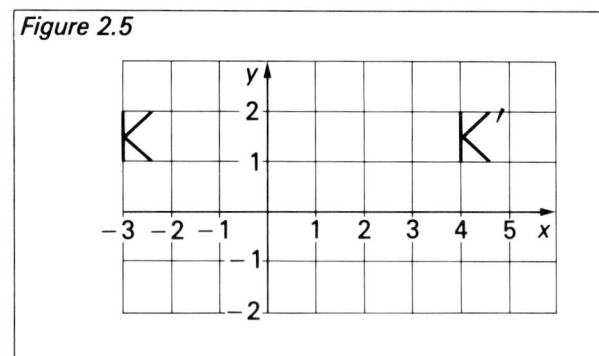

5 What translation has taken V to V′ in Figure 2.6? What two individual translations parallel to the *x*-axis and *y*-axis respectively would replace the single translation from V to V′?

Figure 2.6

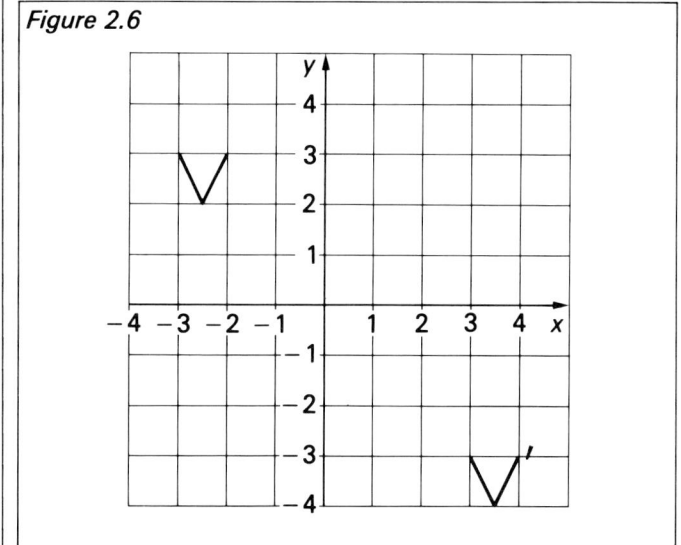

6 Examine Figure 2.7 and answer the questions.
(a) What translation will take tank 1 to tank 2?
(b) What translation will take tank 1 to tank 3?
(c) What translation will take tank 3 to tank 2?

Figure 2.7

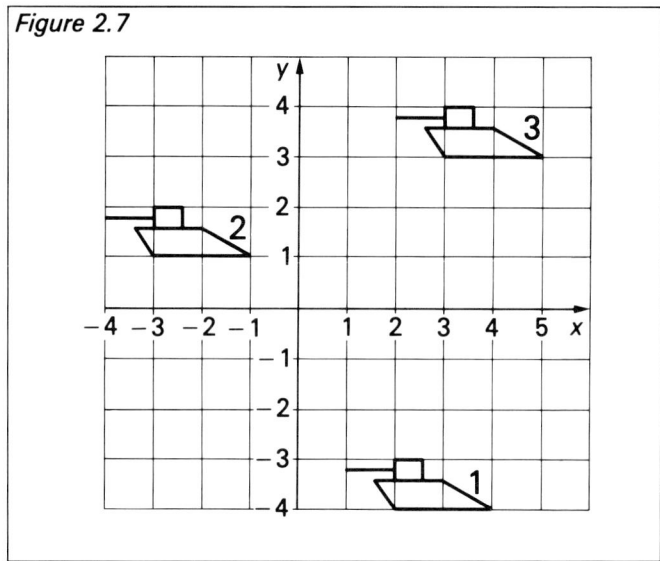

7 Plot the triangle whose vertices have coordinates (−1, 1) (−4, 2) and (−2, 5). Redraw the triangle when it has undergone a translation $\begin{pmatrix} 3 \\ 1 \end{pmatrix}$ (call it triangle 2). Now redraw the second triangle when it has undergone a translation $\begin{pmatrix} -2 \\ -5 \end{pmatrix}$ (call it triangle 3).
    What single translation will take the original triangle to triangle 3?

8 A pinpoint of light on a blackened screen has coordinates of (4, 1). Four successive translations are applied to the pinpoint of light as follows: $\begin{pmatrix} -5 \\ 2 \end{pmatrix}$, $\begin{pmatrix} 8 \\ -1 \end{pmatrix}$, $\begin{pmatrix} 3 \\ -3 \end{pmatrix}$, $\begin{pmatrix} -1 \\ 2 \end{pmatrix}$. If a target has coordinates of $(2, -5)$ what further translation would direct the pinpoint of light to the target?

What single translation would have found the target from the original position of the pinpoint of light?

9 In Figure 2.8 assuming that A remains fixed, what translations must be applied to B and C in order to obtain a square figure of side 2 units?

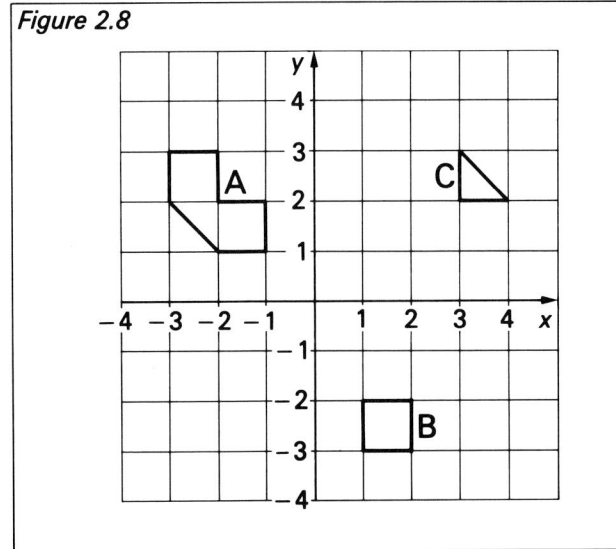

Figure 2.8

10 In Figure 2.9 assuming that B is fixed, what translations must be applied to A, C and D in order to obtain a single large right-angled triangle? (Note: there are several solutions to this problem.)

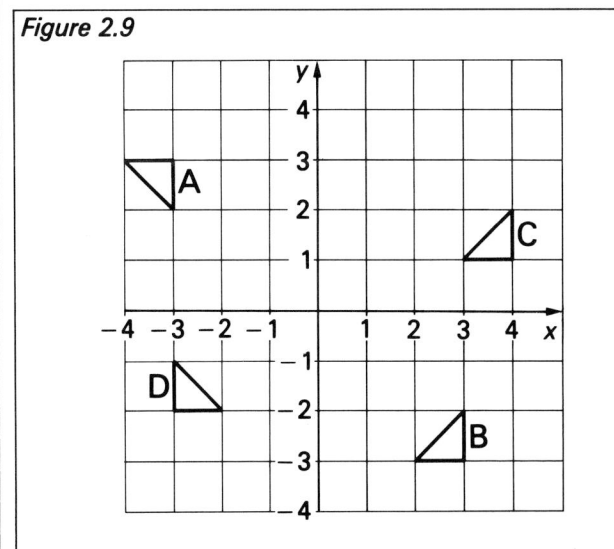

Figure 2.9

Rotations

Figure 2.10 shows a letter F with reference to the cartesian x–y axes.

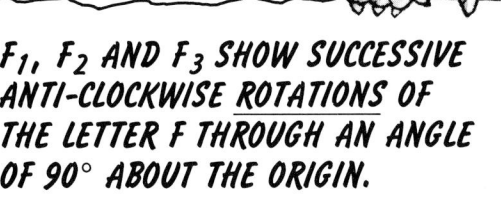

F_1, F_2 AND F_3 SHOW SUCCESSIVE ANTI-CLOCKWISE ROTATIONS OF THE LETTER F THROUGH AN ANGLE OF 90° ABOUT THE ORIGIN.

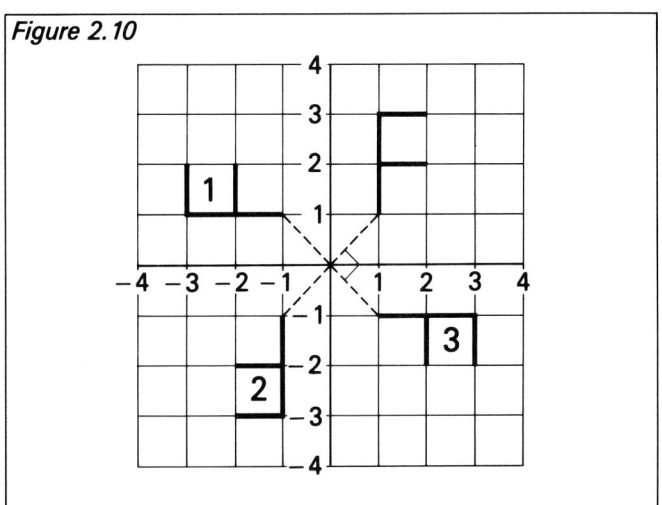

Figure 2.10

Example 2

Copy and complete Figure 2.11 when the triangle (T) undergoes a 90° anti-clockwise rotation about the origin.

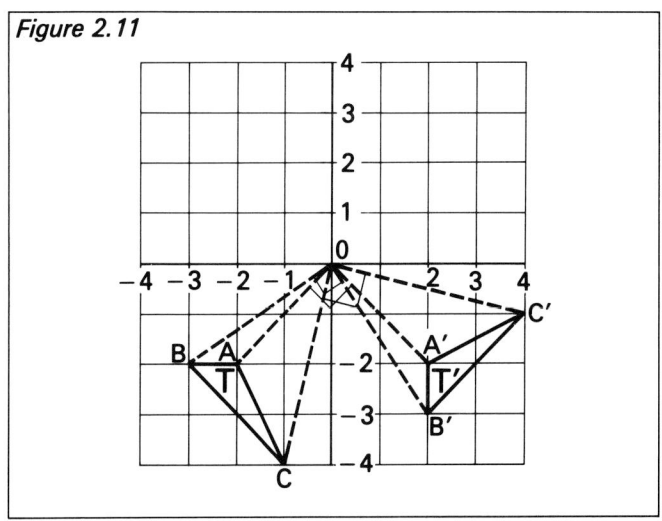

Figure 2.11

Solution

The image of T under the transformation is T'.

$$OA = OA' \text{ and } \angle AOA' = 90°$$
Similarly
$$OB = OB' \text{ and } \angle BOB' = 90°$$
$$OC = OC' \text{ and } \angle COC' = 90°$$

Example 3

What transformation takes place when the matrix $\begin{pmatrix} 0 & -1 \\ 1 & 0 \end{pmatrix}$ is applied to triangle ABC which has vertices with the following coordinates, A = (2, 3), B = (4, 5), C = (6, 4)?

Solution

Plot the points A, B and C and join them together to give triangle ABC (Figure 2.12).

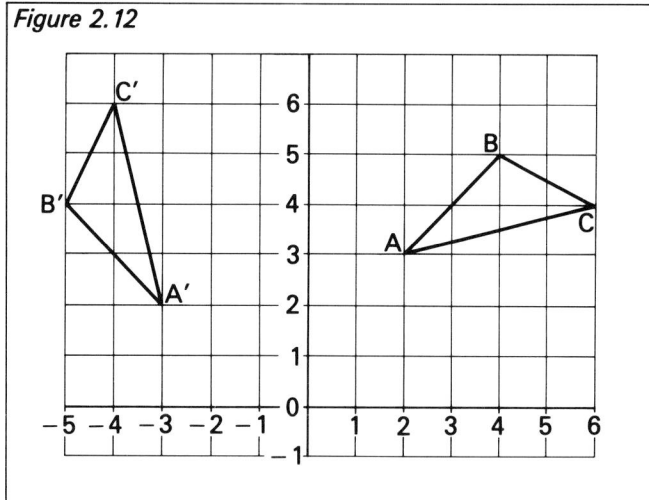

Figure 2.12

Each coordinate pair can be written as a column matrix, thus $\begin{pmatrix} x \\ y \end{pmatrix}$

so $\quad A = \begin{pmatrix} 2 \\ 3 \end{pmatrix}, \quad B = \begin{pmatrix} 4 \\ 5 \end{pmatrix}, \quad C = \begin{pmatrix} 6 \\ 4 \end{pmatrix}.$

Then premultiply each of these matrices individually by the matrix $\begin{pmatrix} 0 & -1 \\ 1 & 0 \end{pmatrix}$ to find their images.

Let the image of A be A'

so $\quad \begin{pmatrix} 0 & -1 \\ 1 & 0 \end{pmatrix}\begin{pmatrix} 2 \\ 3 \end{pmatrix} = \begin{pmatrix} -3 \\ 2 \end{pmatrix}$

hence A' = (−3, 2).
Let the image of B be B'

$$\begin{pmatrix} 0 & -1 \\ 1 & 0 \end{pmatrix}\begin{pmatrix} 4 \\ 5 \end{pmatrix} = \begin{pmatrix} -5 \\ 4 \end{pmatrix}$$

hence B' = (−5, 4).
Let the image of C be C'

$$\begin{pmatrix} 0 & -1 \\ 1 & 0 \end{pmatrix}\begin{pmatrix} 6 \\ 4 \end{pmatrix} = \begin{pmatrix} -4 \\ 6 \end{pmatrix}$$

hence C' = (−4, 6).

Plotting the new coordinates A', B' and C' and completing the triangle (Figure 2.12), it can be seen that the transformation is a 90° rotation in an anti-clockwise direction about the origin.

Exercise 2.2

1 Copy Figure 2.13 and draw in the position of the shape when it is rotated through 90° in an anti-clockwise direction.

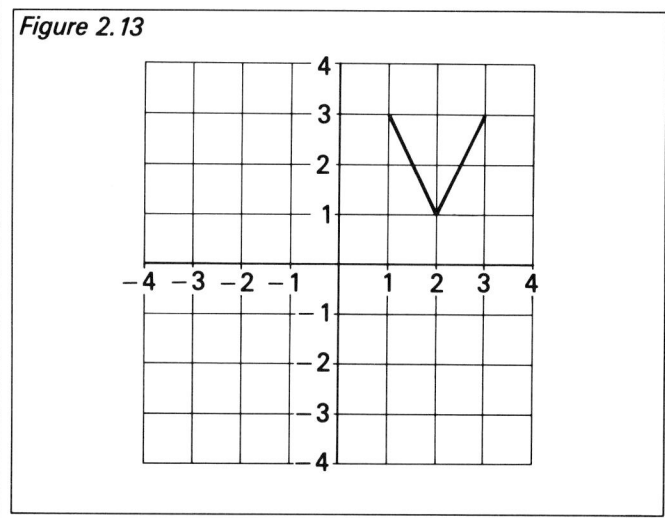

Figure 2.13

2 Copy Figure 2.14 and draw in the position of the letter P when it has been rotated through 180° about the origin.

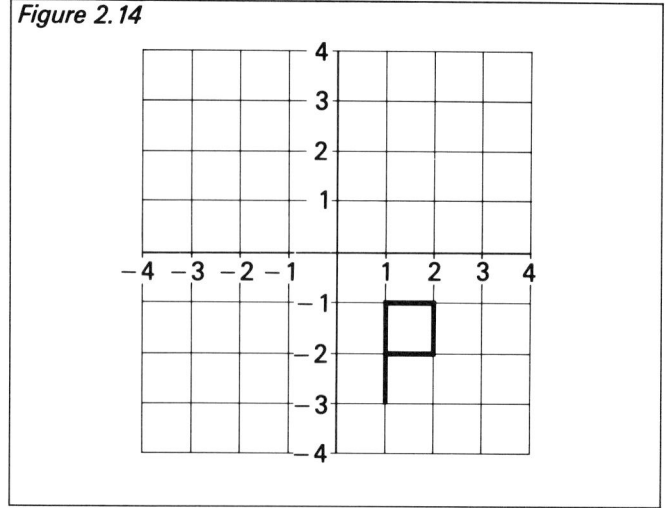

Figure 2.14

3 The arrowhead shown in Figure 2.15 is subjected to a clockwise rotation of 90° about the origin. Draw in its new position.

Figure 2.15

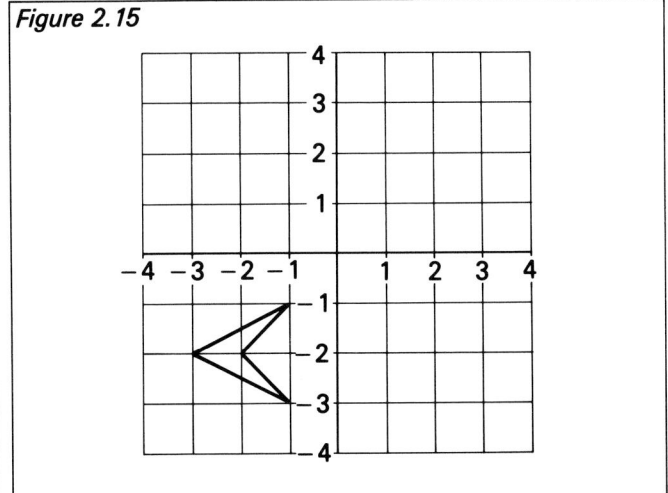

4 In Figure 2.16 the letter T is rotated by 270° in an anti-clockwise direction about the origin. Draw in its new position.

Figure 2.16

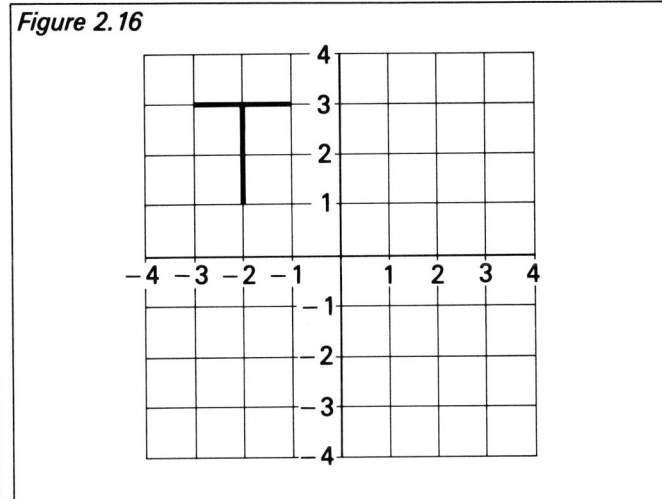

5 Redraw the shape given in Figure 2.17 when it is rotated by 270° in a clockwise direction about the origin.

6 Draw in the new position of the letter Y in Figure 2.18 when it is rotated by 90° in an anti-clockwise direction about the point (1, 1).

7 Draw in the new position when the shape in Figure 2.19 is rotated by 180° about the point (1, −2).

8 A triangle XYZ has the coordinates of its vertices as X = (−1, −1), Y = (−2, −3) and Z = (−3, −2). It is then subjected to a transformation which moves the triangle to X'Y'Z', where X' = (−1, 1), Y' = (−3, 2) and Z' = (−2, 3). By first plotting the two triangles on a graph describe which transformation has taken place.

Figure 2.17

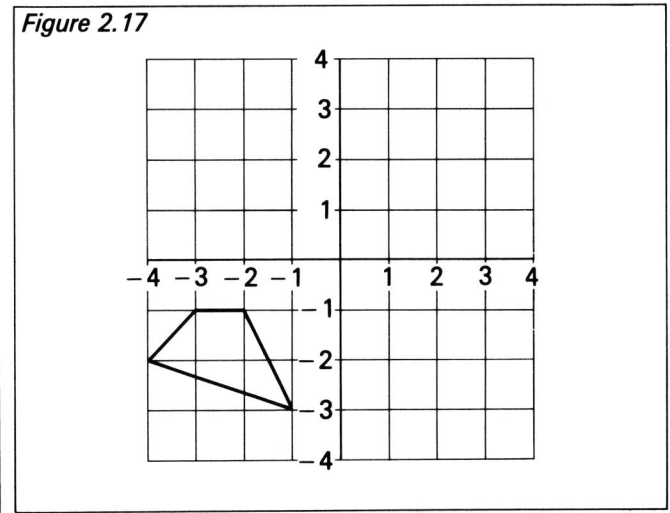

Figure 2.18

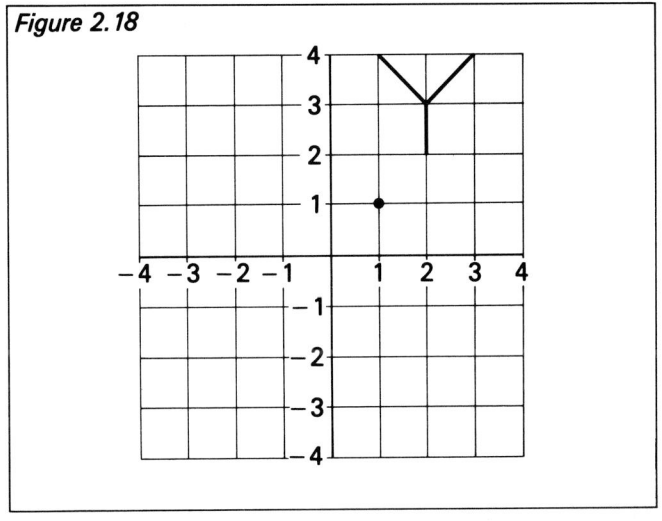

Figure 2.19

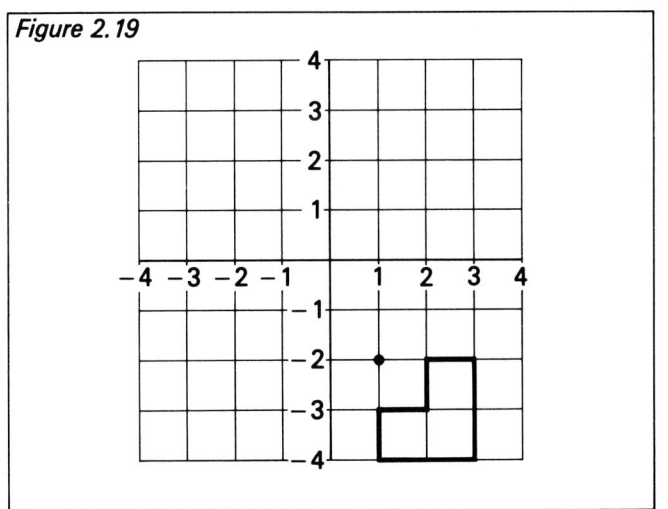

9 Copy the shape given in Figure 2.20 and then redraw it on the same axes when it has undergone a 270° clockwise rotation about the origin.

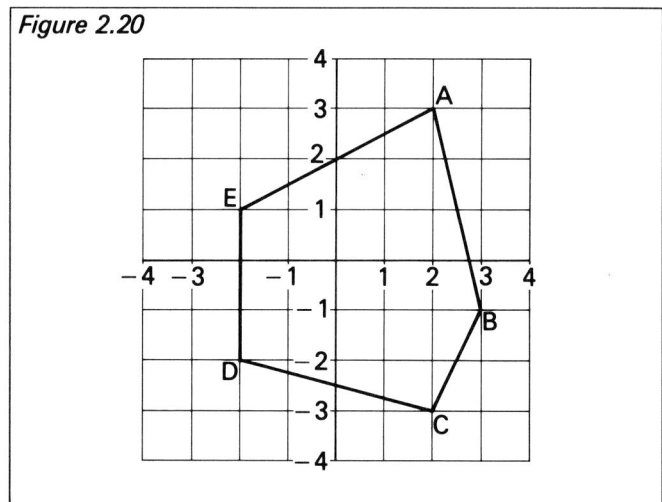

Figure 2.20

10 Examine Figure 2.21 and describe exactly what transformation has taken place.

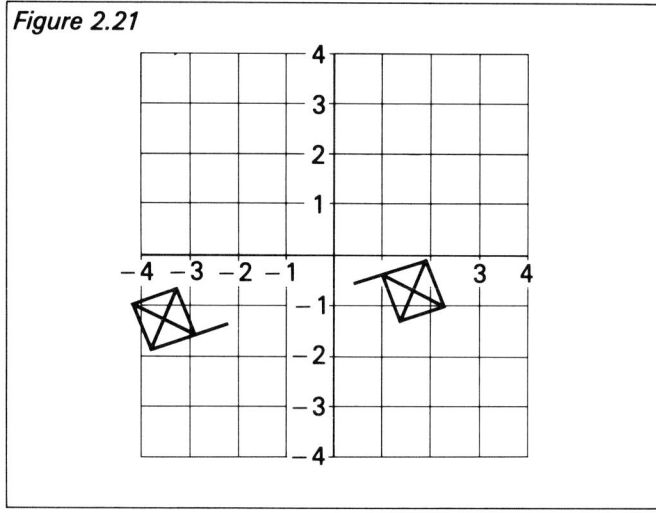

Figure 2.21

11 Plot the triangle CDE where C = (−3, −1), D = (−1, −2) and E = (−2, −3). Rewrite each coordinate as a column matrix and premultiply each one by the matrix $\begin{pmatrix} -1 & 0 \\ 0 & -1 \end{pmatrix}$. Draw the image triangle C′D′E′ and state what transformation has taken place.

12 The vertices of a kite have the following coordinates L = (3, 2), M = (4, 1), N = (3, −1) and O = (2, 1).

Plot this shape and then apply the matrix $\begin{pmatrix} 0 & -1 \\ 1 & 0 \end{pmatrix}$ to each coordinate pair. Redraw the transformation stating carefully what it is.

13 An irregular pentagon is represented by the coordinates PQRST where P = (−3, −1), Q = (−2, 2), R = (2, 3), S = (3, 0) and T = (2, −2). Plot the points and draw the shape. Now apply the matrix $\begin{pmatrix} 0 & 1 \\ -1 & 0 \end{pmatrix}$ to this shape redrawing the transformation on the same graph. What is the transformation?

14 Copy Figure 2.22 and apply the matrix $\begin{pmatrix} -1 & 0 \\ 0 & -1 \end{pmatrix}$ to the shape. Redraw and name the transformation.

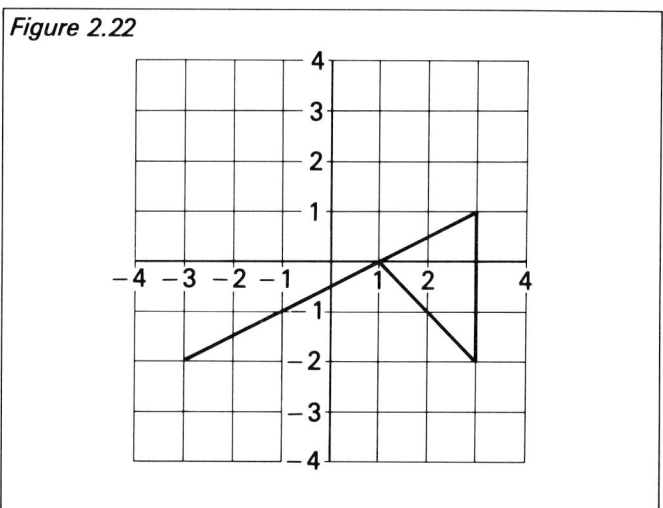

Figure 2.22

15 Apply the matrix $\begin{pmatrix} 0 & 1 \\ -1 & 0 \end{pmatrix}$ to the parallelogram given in Figure 2.23. Redraw and name the transformation.

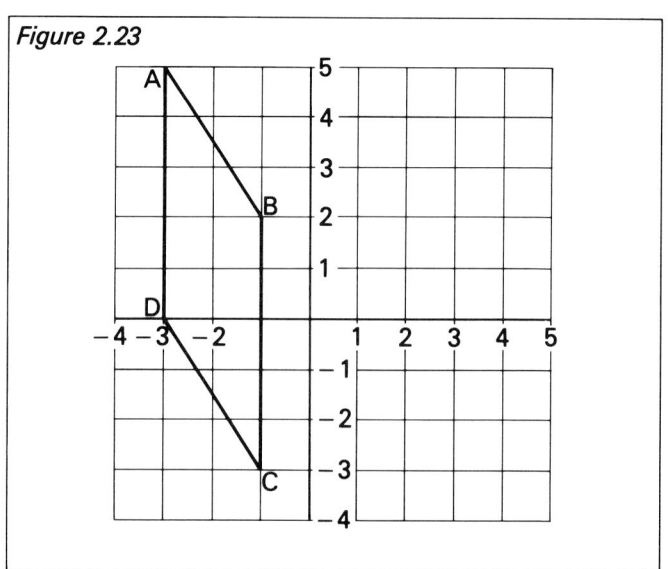

Figure 2.23

Reflections

Figure 2.24 shows a letter K with reference to the cartesian x–y axes.

Figure 2.24

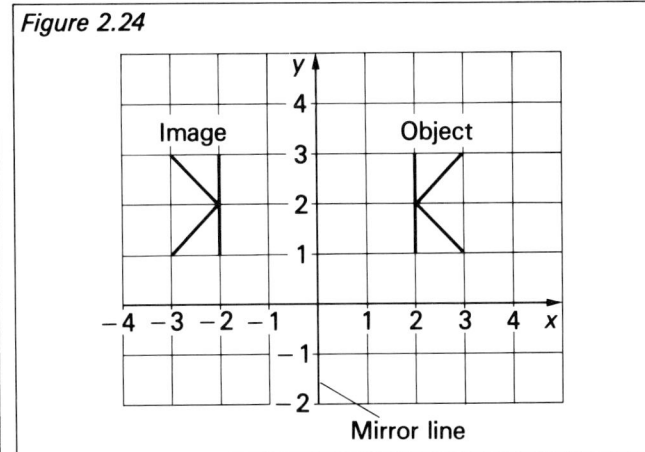

IF THE y-AXIS IS A MIRROR LINE THEN THE <u>REFLECTION</u> (OR IMAGE) OF K IS SHOWN ON THE LEFT OF THE MIRROR LINE.

Any point on the image is exactly the same perpendicular distance from the mirror line as the corresponding point on the object.

Example 4

Draw in the reflection of triangle ABC in Figure 2.25 if the mirror line has the equation $y = x$.

Solution

The image of triangle ABC is triangle A'B'C'.

AM = A'M
BN = B'N
CO = C'O

All these lines are perpendicular to the mirror line $y = x$.

Figure 2.25

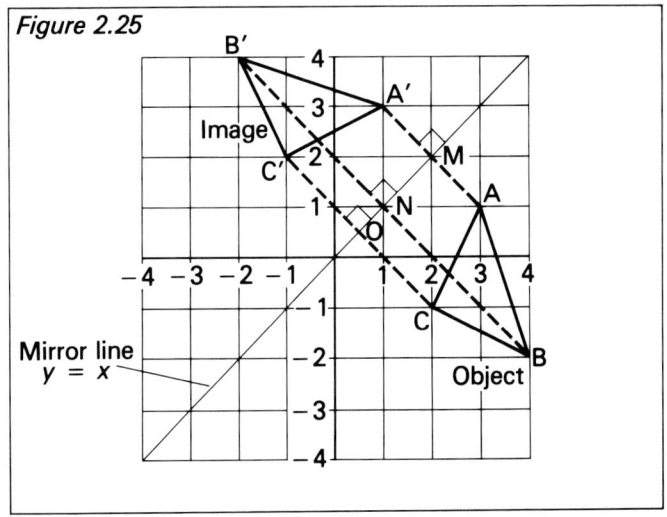

Example 5

In Figure 2.26 triangle ABC has coordinates A = (2, 1), B = (1, 2) and C = (4, 3). The matrix $\begin{pmatrix} 1 & 0 \\ 0 & -1 \end{pmatrix}$ is applied to the triangle. Redraw the triangle once this transformation has taken place and describe what has happened.

Figure 2.26

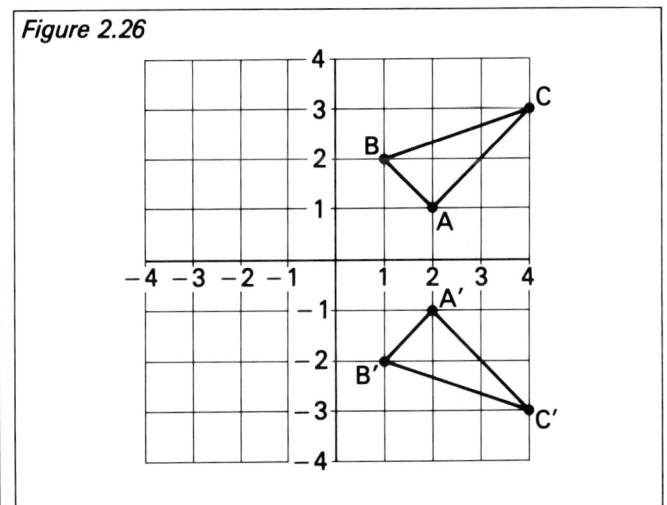

Solution

Write each coordinate as a column matrix, thus $\begin{pmatrix} x \\ y \end{pmatrix}$.

$$A = \begin{pmatrix} 2 \\ 1 \end{pmatrix}, \quad B = \begin{pmatrix} 1 \\ 2 \end{pmatrix}, \quad C = \begin{pmatrix} 4 \\ 3 \end{pmatrix}.$$

Then premultiply each coordinate individually by the matrix $\begin{pmatrix} 1 & 0 \\ 0 & -1 \end{pmatrix}$.

Let A' be the image of A

19

so $\begin{pmatrix} 1 & 0 \\ 0 & -1 \end{pmatrix}\begin{pmatrix} 2 \\ 1 \end{pmatrix} = \begin{pmatrix} 2 \\ -1 \end{pmatrix}$ and A′ = (2, −1).

Let B′ be the image of B

so $\begin{pmatrix} 1 & 0 \\ 0 & -1 \end{pmatrix}\begin{pmatrix} 1 \\ 2 \end{pmatrix} = \begin{pmatrix} 1 \\ -2 \end{pmatrix}$ and B′ = (1, −2).

Let C′ be the image of C

so $\begin{pmatrix} 1 & 0 \\ 0 & -1 \end{pmatrix}\begin{pmatrix} 4 \\ 3 \end{pmatrix} = \begin{pmatrix} 4 \\ -3 \end{pmatrix}$ and C′ = (4, −3).

By plotting the new triangle A′B′C′, it will be seen that the transformation is a reflection in the x-axis (the x-axis is the mirror line).

Exercise 2.3

1 Copy Figure 2.27 and draw in the reflection of the letter A if the x-axis is the mirror line.

Figure 2.27

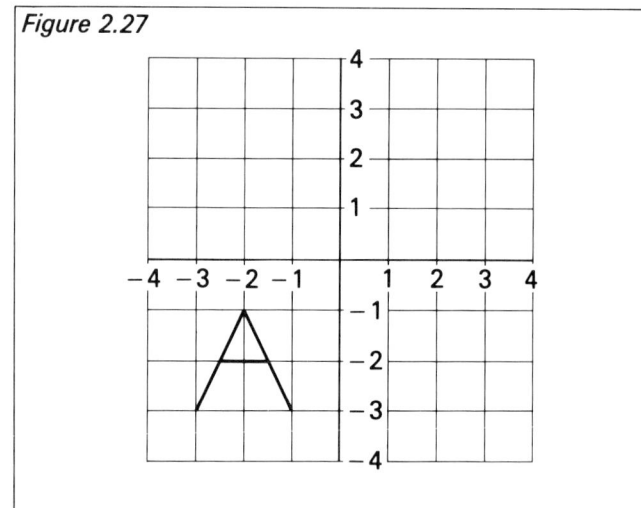

2 Copy Figure 2.28 and using the y-axis as the mirror line draw in the reflection of the letter E.

Figure 2.28

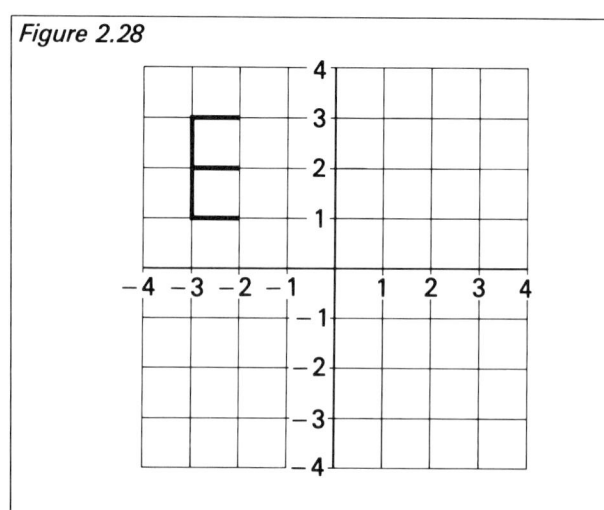

3 Draw in the reflection of the shape given in Figure 2.29 if the mirror line has the equation $y = -x$.

Figure 2.29

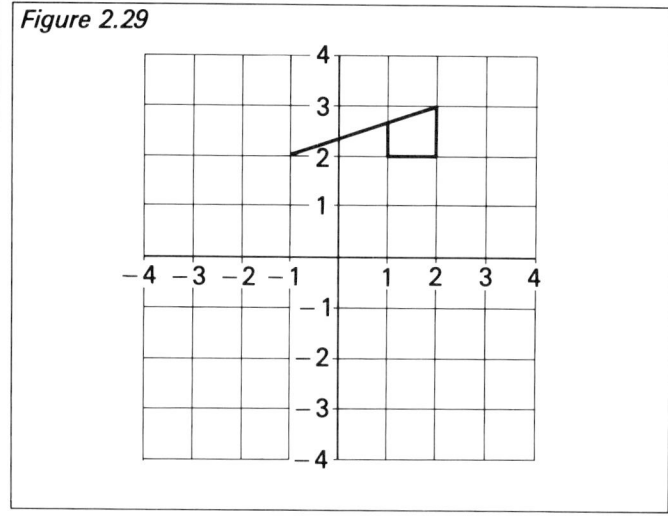

4 Copy and complete the shape given in Figure 2.30 and draw in its image if the mirror line has the equation $y = -1$.

Figure 2.30

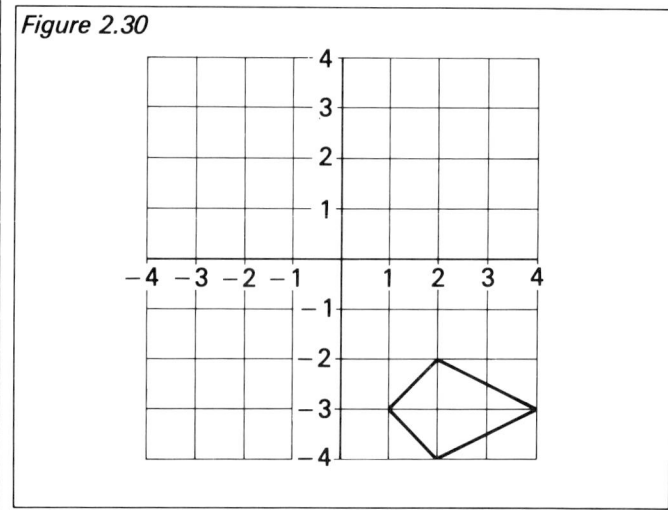

5 Draw the reflection of the shape, given in Figure 2.31, in the mirror line $x = -2$.

6 If the mirror line is given by the equation $x = 0$, copy Figure 2.32 and then draw the image of the given quadrilateral.

7 Draw in the mirror line in Figure 2.33 and give its equation.

8 Draw in the mirror line in Figure 2.34 and give its equation.

Figure 2.31

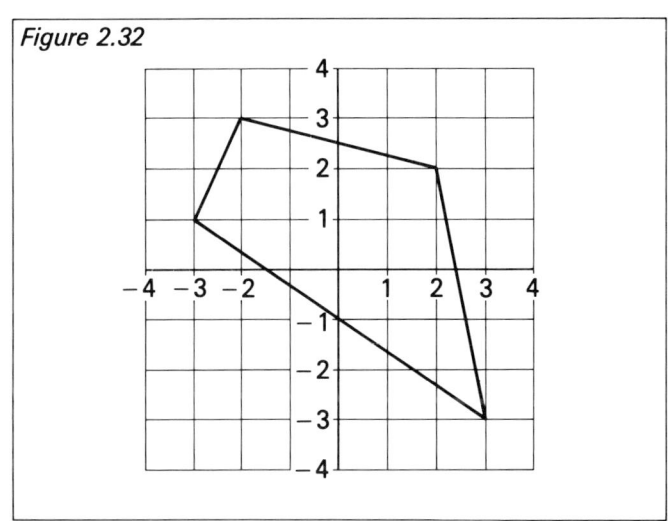

Figure 2.32

Figure 2.33

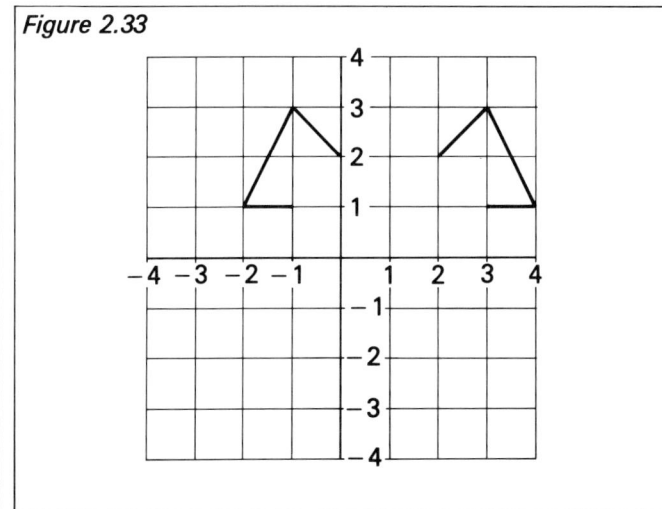

Figure 2.34

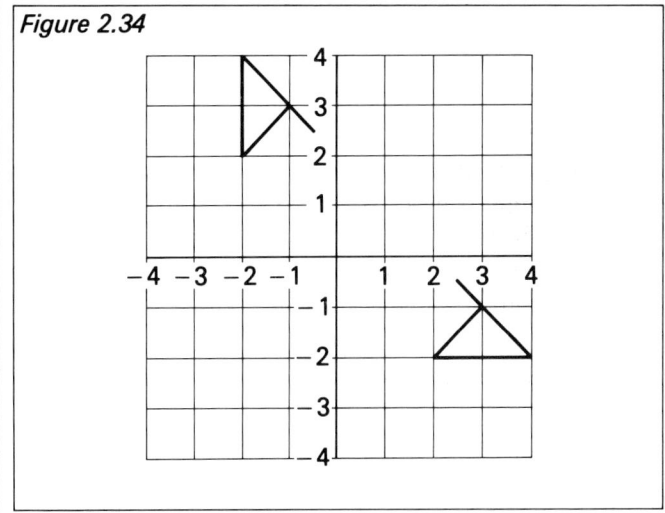

9 Draw the shape given in Figure 2.35 and using the mirror line $y = -x$ complete the image.

Figure 2.35

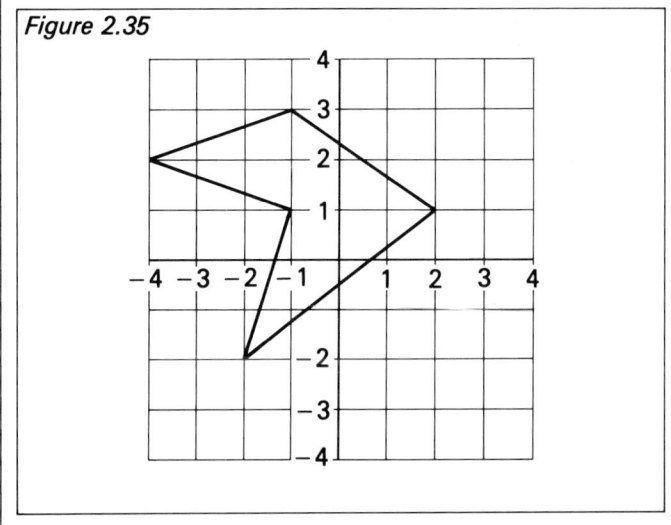

10 A triangle PQR has coordinates $P = (-2, -1)$, $Q = (-3, -4)$ and $R = (-1, -3)$. Rewrite each coordinate pair as a column matrix and then premultiply each one individually by the matrix $\begin{pmatrix} -1 & 0 \\ 0 & 1 \end{pmatrix}$. If P′, Q′ and R′ are the images of P, Q and R respectively then draw triangle P′Q′R′ stating what transformation has taken place.

11 The vertices of a parallelogram ABCD are represented by the coordinates $A = (2, -1)$, $B = (4, -1)$, $C = (2, -3)$ and $D = (0, -3)$. Plot and draw the parallelogram and then apply the transformation matrix $\begin{pmatrix} 0 & -1 \\ -1 & 0 \end{pmatrix}$. Draw the transformation and state what has happened.

21

12 A trapezium WXYZ has vertices with the following coordinates W = (−1, 3), X = (2, 3), Y = (−2, −1) and Z = (−3, 1). By applying the matrix $\begin{pmatrix} 1 & 0 \\ 0 & -1 \end{pmatrix}$ to each coordinate pair, draw the new transformation and name it.

13 Copy the diagram given in Figure 2.36 and apply the matrix $\begin{pmatrix} -1 & 0 \\ 0 & 1 \end{pmatrix}$. Draw and name the new transformation.

Figure 2.36

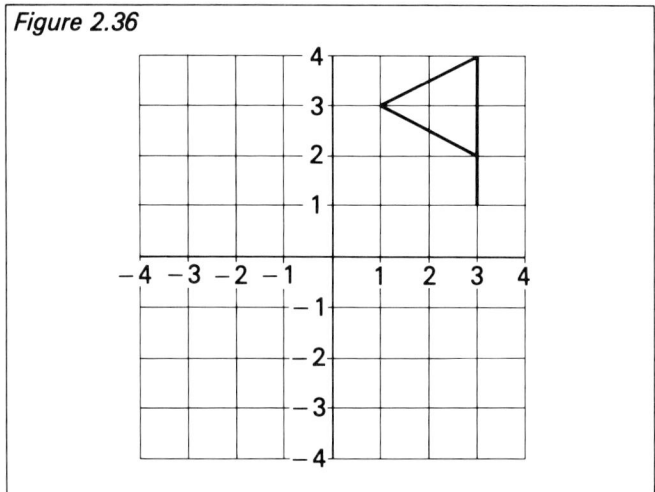

14 Apply the transformation $\begin{pmatrix} 0 & -1 \\ -1 & 0 \end{pmatrix}$ to the letter V in Figure 2.37. Draw and name the new transformation.

Figure 2.37

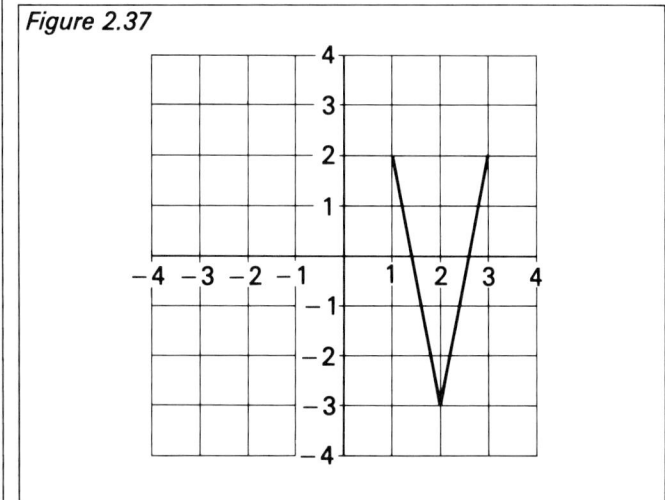

15 Plot the quadrilateral KLMN which has the following coordinates: K = (−1, 2), L = (2, 1), M = (0, −3), N = (−3, −3). When the matrix $\begin{pmatrix} 1 & 0 \\ 0 & -1 \end{pmatrix}$ is applied, a transformation takes place. Redraw the quadrilateral, calling its image K′L′M′N′. What is the name of the transformation that has taken place?

Enlargements

IN AN _ENLARGEMENT_ TRANSFORMATION, THE SHAPE OF A FIGURE REMAINS THE SAME BUT THE LENGTHS OF ALL ITS SIDES ARE INCREASED OR DECREASED BY A CONSTANT FACTOR.

Figure 2.38

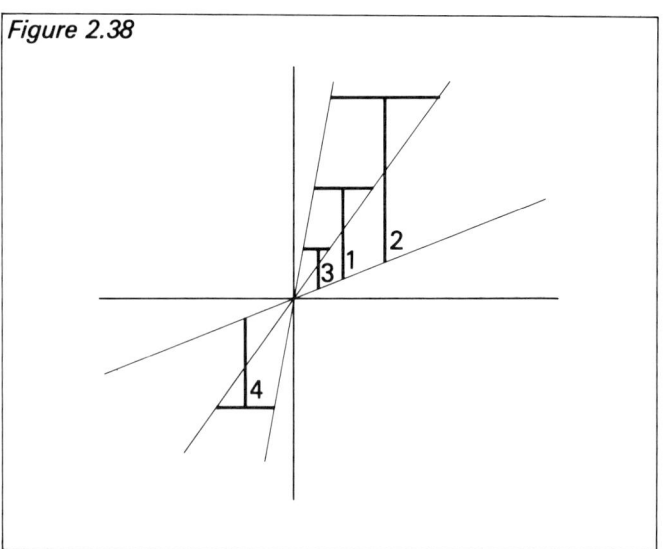

In Figure 2.38 the origin is the centre of enlargement.
   T_1 is the original shape.
   T_2 is an enlargement by a scale factor of 2 (this means that every length is 2 times as long as the original).

T_3 is an enlargement by a scale factor of $\frac{1}{2}$. (Every length is $\frac{1}{2}$ that of the original.)

T_4 is an enlargement by a scale factor of -1. (Every length is the same as the original but the figure undergoes a 180° rotation about the centre of enlargement.)

Example 6

Draw the enlargement of the square ABCD in Figure 2.39 with a scale factor of 3.

Solution

Draw in the lines OA, OB, OC and OD making sure that they are extended to the edge of the graph. Measure OA and multiply this value by 3 since this is the enlargement factor. Mark off the point A′ such that OA′ = 3OA. Repeat this procedure with OB, OC and OD marking off B′, C′ and D′ respectively. By joining together the points A′, B′, C′ and D′ the enlarged square is obtained with all sides increased by a scale factor of 3.

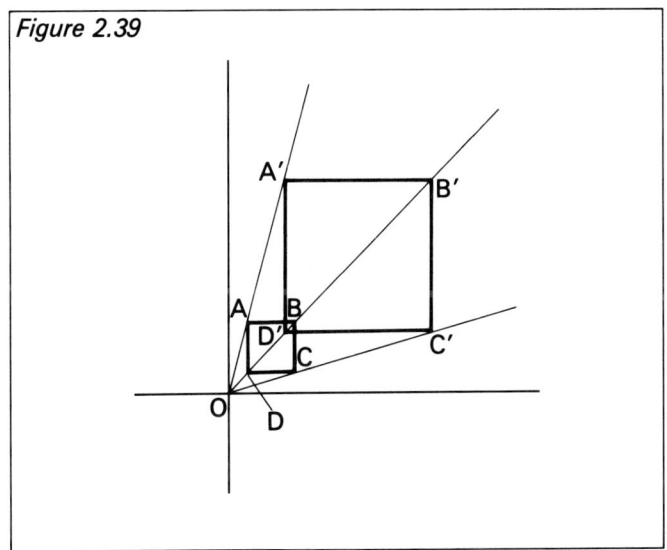

Figure 2.39

ENLARGEMENTS CAN ALSO BE MADE BY APPLYING THE MATRIX

$$\begin{pmatrix} n & 0 \\ 0 & n \end{pmatrix}$$

TO ANY SHAPE. THE VALUE OF n IS THE SCALE FACTOR OF THE ENLARGEMENT.

Example 7

Apply matrix $\begin{pmatrix} 2 & 0 \\ 0 & 2 \end{pmatrix}$ to the vertices of the triangle ABC when A = (2, 3), B = (4, 5) and C = (6, 2). What transformation takes place? (See Figure 2.40.)

Solution

Write each coordinate pair as a column matrix, thus $\begin{pmatrix} x \\ y \end{pmatrix}$.

$$A = \begin{pmatrix} 2 \\ 3 \end{pmatrix}, \quad B = \begin{pmatrix} 4 \\ 5 \end{pmatrix}, \quad C = \begin{pmatrix} 6 \\ 2 \end{pmatrix}.$$

Then premultiply each column matrix by the matrix $\begin{pmatrix} 2 & 0 \\ 0 & 2 \end{pmatrix}$.

Let the image of A be A′

so $\begin{pmatrix} 2 & 0 \\ 0 & 2 \end{pmatrix}\begin{pmatrix} 2 \\ 3 \end{pmatrix} = \begin{pmatrix} 4 \\ 6 \end{pmatrix}$ and A′ = (4, 6).

Let the image of B be B′

so $\begin{pmatrix} 2 & 0 \\ 0 & 2 \end{pmatrix}\begin{pmatrix} 4 \\ 5 \end{pmatrix} = \begin{pmatrix} 8 \\ 10 \end{pmatrix}$ and B′ = (8, 10).

Let the image of C be C′

so $\begin{pmatrix} 2 & 0 \\ 0 & 2 \end{pmatrix}\begin{pmatrix} 6 \\ 2 \end{pmatrix} = \begin{pmatrix} 12 \\ 4 \end{pmatrix}$ and C′ = (12, 4).

By plotting the transformed triangle A′B′C′, it will be seen that its sides have been enlarged by a factor of 2.

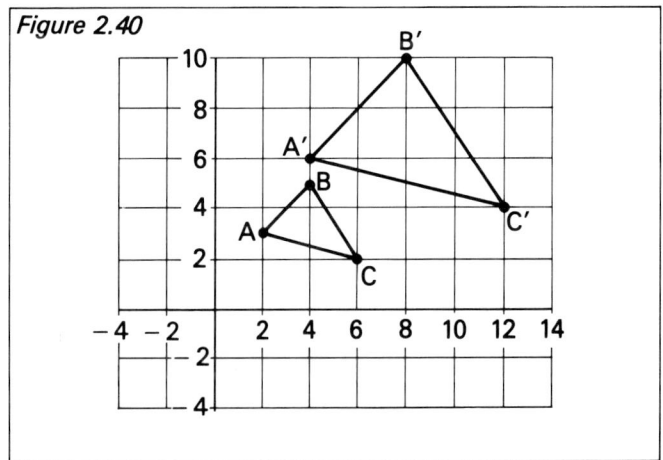

Figure 2.40

Exercise 2.4

1 Enlarge the triangle in Figure 2.41 by a factor of 2 with the origin as the centre of enlargement.

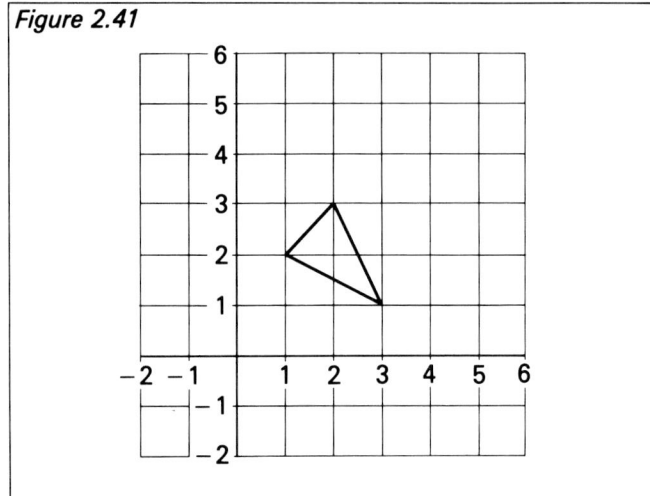

Figure 2.41

2 Enlarge the quadrilateral in Figure 2.42 by a factor of $\frac{1}{2}$ with the origin as the centre of enlargement.

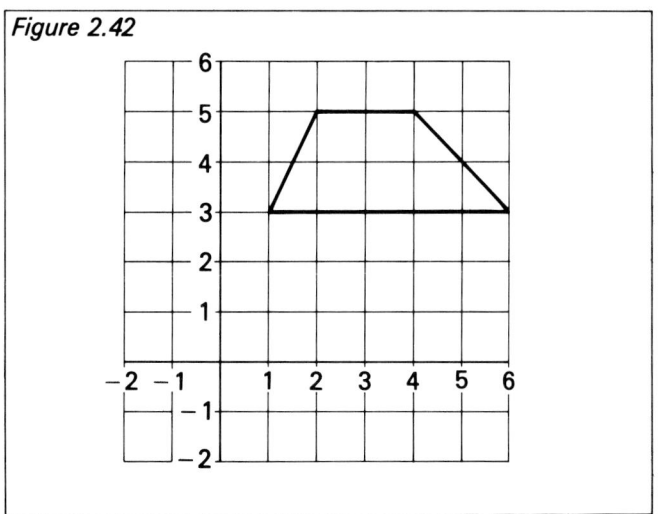

Figure 2.42

3 Enlarge the shape in Figure 2.43 by a factor of -2 with the origin as the centre of enlargement.

4 Enlarge the Z shape in Figure 2.44 by a scale factor of 3 with the origin as the centre of enlargement.

5 Enlarge the shape in Figure 2.45 by a scale factor of -1 with the origin as the centre of enlargement.

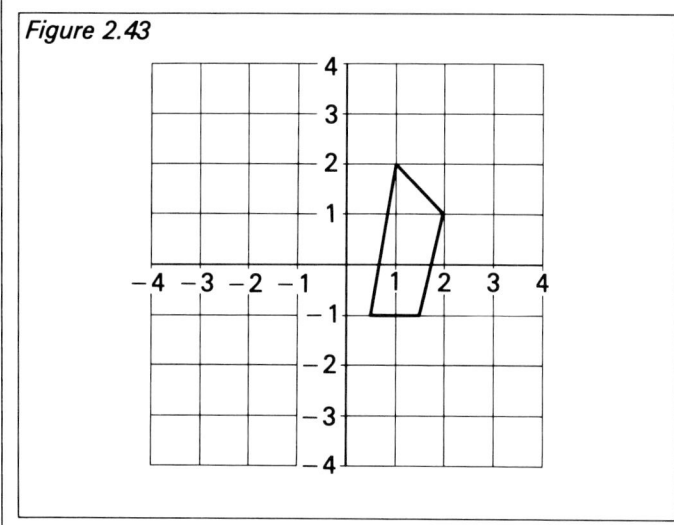

Figure 2.43

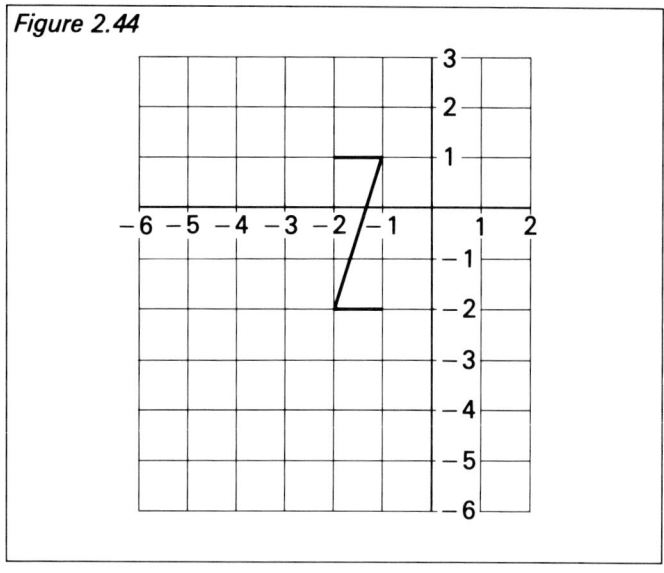

Figure 2.44

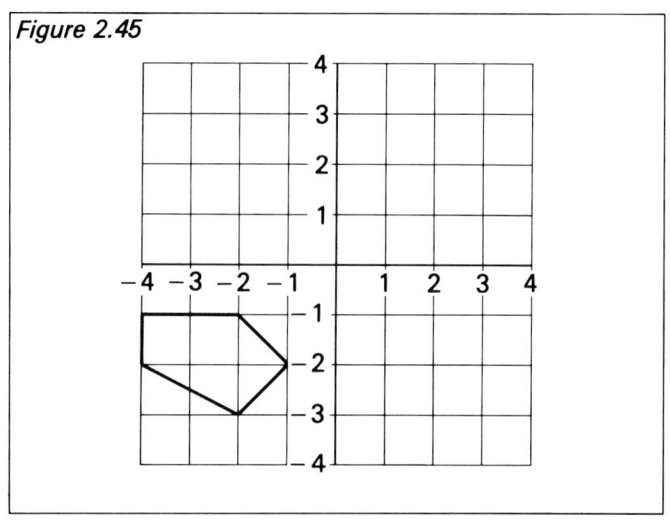

Figure 2.45

6 Enlarge the trapezium in Figure 2.46 by a scale factor of $-\frac{1}{2}$ with the origin as the centre of enlargement.

Figure 2.46

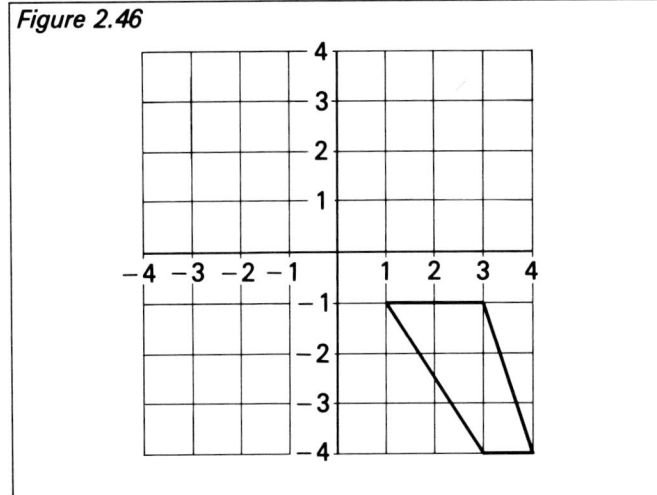

7 Enlarge the shape in Figure 2.47 by a scale factor of $\frac{1}{3}$ using the origin as the centre of enlargement.

Figure 2.47

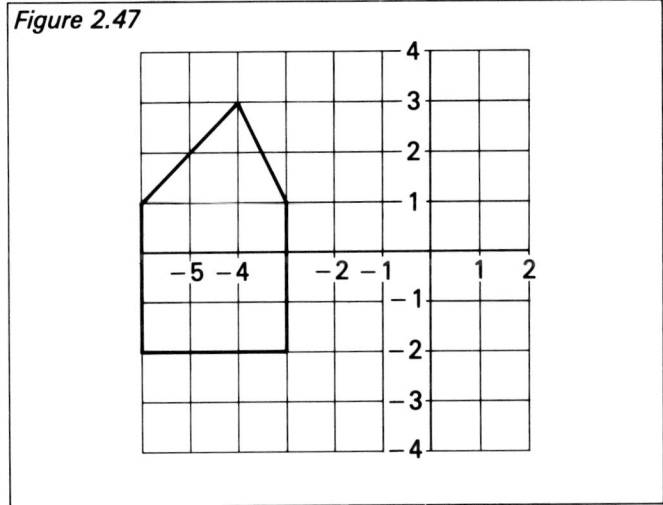

8 Enlarge the quadrilateral in Figure 2.48 by a scale factor of -3 with the origin as the centre of enlargement.

9 Enlarge the shape in Figure 2.49 by a scale factor of $1\frac{1}{2}$ using the origin as the centre of enlargement.

10 Enlarge the trapezium in Figure 2.50 by a scale factor of $-\frac{1}{2}$ using the origin as the centre of enlargement.

Figure 2.48

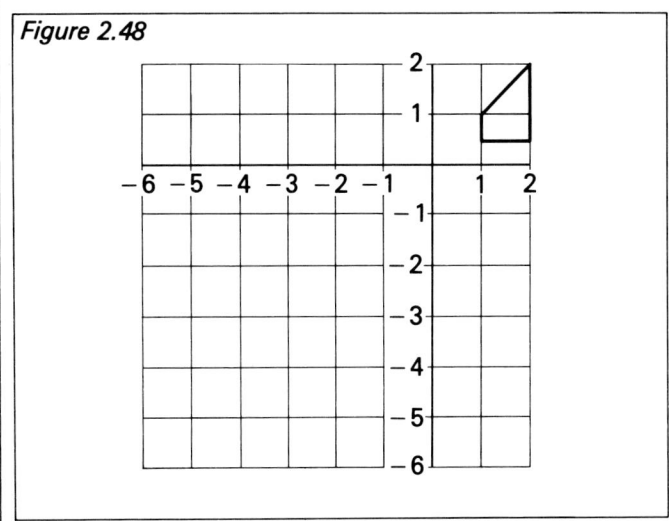

Figure 2.49

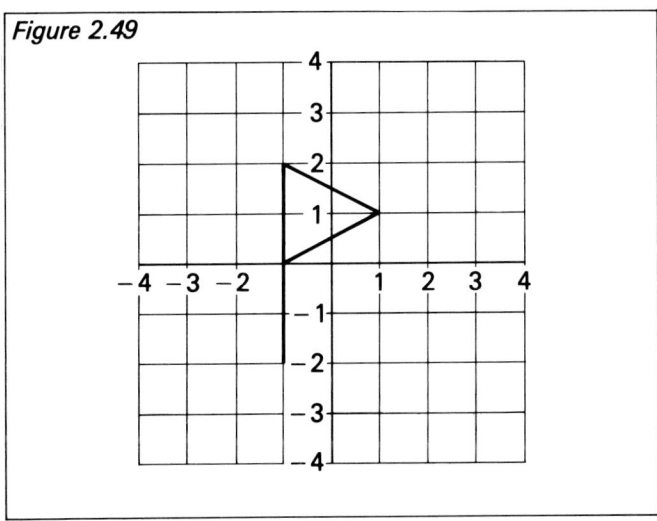

Figure 2.50

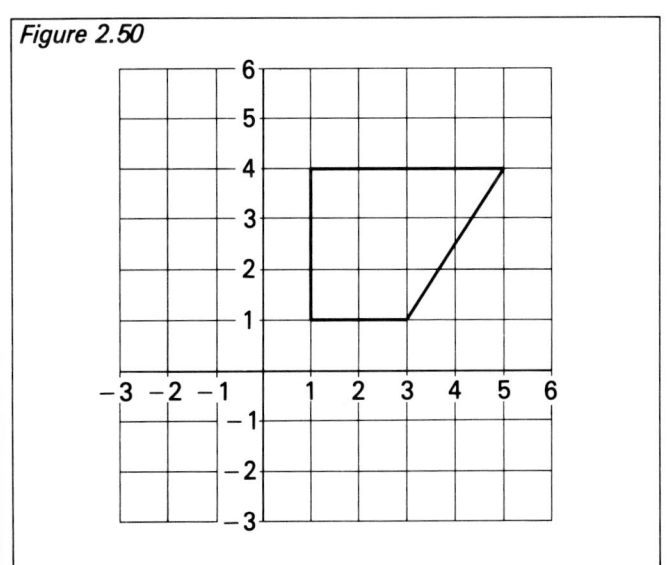

11 A triangle ABC has vertices with the following coordinates: A = (−1, −1), B = (1, 2), C = (2, 0). Rewrite these coordinate pairs as column matrices and premultiply each by the matrix $\begin{pmatrix} 3 & 0 \\ 0 & 3 \end{pmatrix}$. Plot triangle ABC and its transformation. State what kind of transformation has taken place.

12 The vertices of a trapezium PQRS are given as P = (−4, −2), Q = (−3, 3), R = (0, 3) and S = (2, −2). Plot and draw the trapezium. On the same graph draw in and state the transformation when the matrix $\begin{pmatrix} -\frac{1}{2} & 0 \\ 0 & -\frac{1}{2} \end{pmatrix}$ is applied to the vertices of the trapezium.

13 A quadrilateral WXYZ has the following coordinates: W = (1, −1), X = (4, −2), Y = (3, −3) and Z = (2, −3). Plot this figure and then redraw and name the transformation when matrix $\begin{pmatrix} -2 & 0 \\ 0 & -2 \end{pmatrix}$ has been applied to the vertices.

14 Copy Figure 2.51. Now apply the matrix $\begin{pmatrix} 2 & 0 \\ 0 & 2 \end{pmatrix}$ to four selected coordinate pairs on the figure. Redraw the transformation on the same graph and state what has happened.

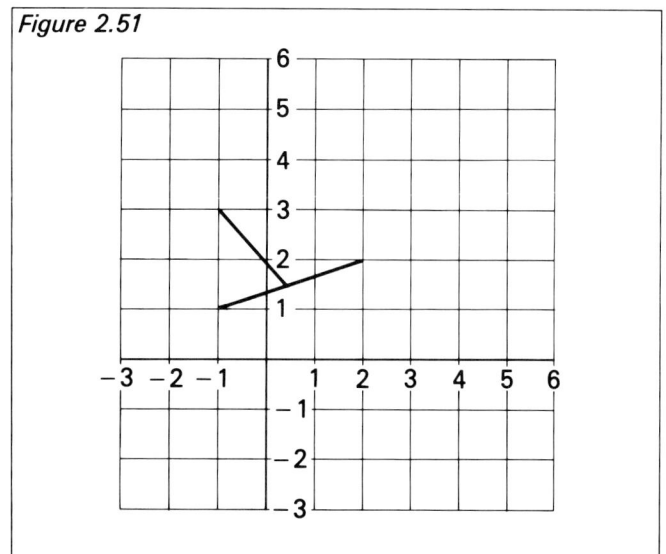

Figure 2.51

15 Copy Figure 2.52 and apply the matrix $\begin{pmatrix} -1\frac{1}{2} & 0 \\ 0 & -1\frac{1}{2} \end{pmatrix}$. Redraw. What is the transformation?

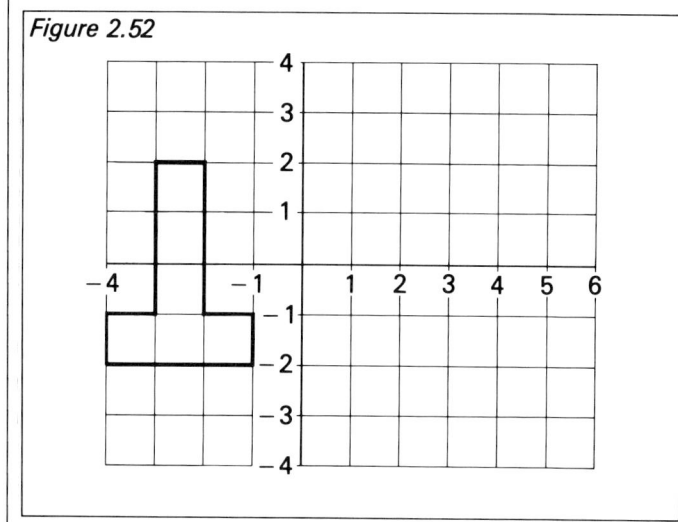

Figure 2.52

Combined transformations

TRANSFORMATIONS CAN BE <u>COMBINED</u> IN VARIOUS WAYS.

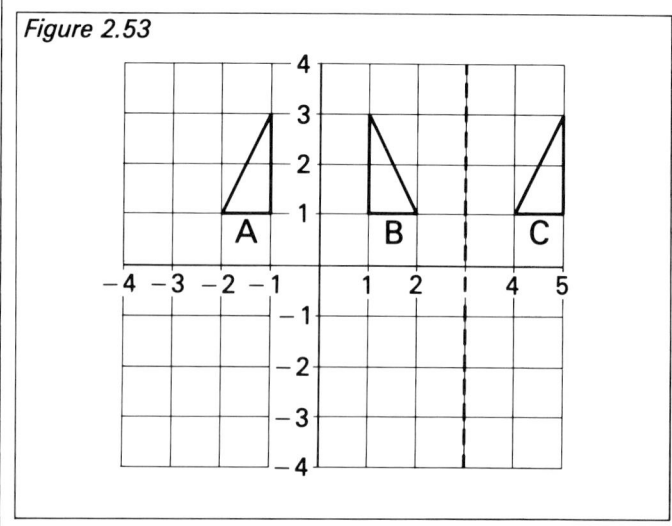

Figure 2.53

In Figure 2.53 triangle A can be reflected in the mirror line $x = 0$ to give triangle B.

Triangle B can be further reflected in mirror line $x = 3$ to give triangle C.

These two reflections could have been replaced by the single translation $\begin{pmatrix} 6 \\ 0 \end{pmatrix}$ taking triangle A directly to triangle C.

Example 8

A letter V has coordinates of (1, 3), (2, 1) and (3, 3) (see Figure 2.54). Apply the matrix $\begin{pmatrix} 0 & 2 \\ -2 & 0 \end{pmatrix}$ to each coordinate pair in turn and redraw the figure. What two transformations have taken place?

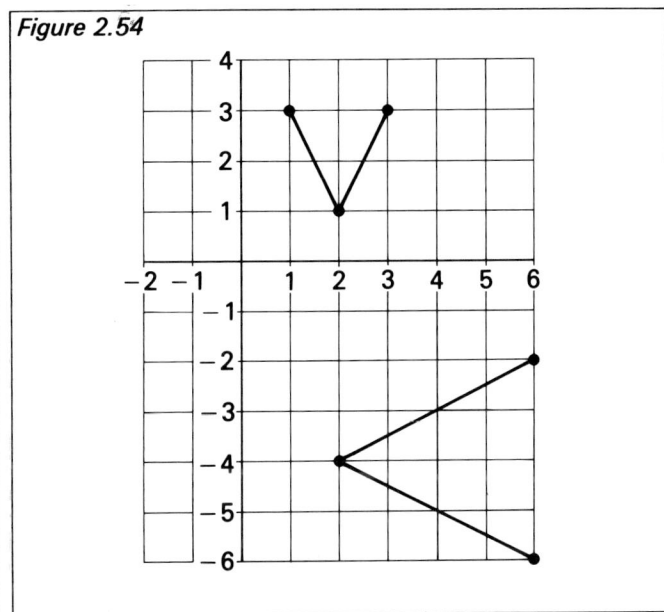

Figure 2.54

Solution

Premultiply each coordinate pair by $\begin{pmatrix} 0 & 2 \\ -2 & 0 \end{pmatrix}$ after they have first been rewritten as column matrices. Thus

$$\begin{pmatrix} 0 & 2 \\ -2 & 0 \end{pmatrix} \begin{pmatrix} 1 \\ 3 \end{pmatrix} = \begin{pmatrix} 6 \\ -2 \end{pmatrix},$$

$$\begin{pmatrix} 0 & 2 \\ -2 & 0 \end{pmatrix} \begin{pmatrix} 2 \\ 1 \end{pmatrix} = \begin{pmatrix} 2 \\ -4 \end{pmatrix},$$

$$\begin{pmatrix} 0 & 2 \\ -2 & 0 \end{pmatrix} \begin{pmatrix} 3 \\ 3 \end{pmatrix} = \begin{pmatrix} 6 \\ -6 \end{pmatrix}.$$

The two transformations are a 270° anti-clockwise rotation about the origin and an enlargement by a scale factor of 2.

Exercise 2.5

1 Reflect the letter T in Figure 2.55 in the x-axis followed by a reflection in the y-axis. What single transformation can replace these two?

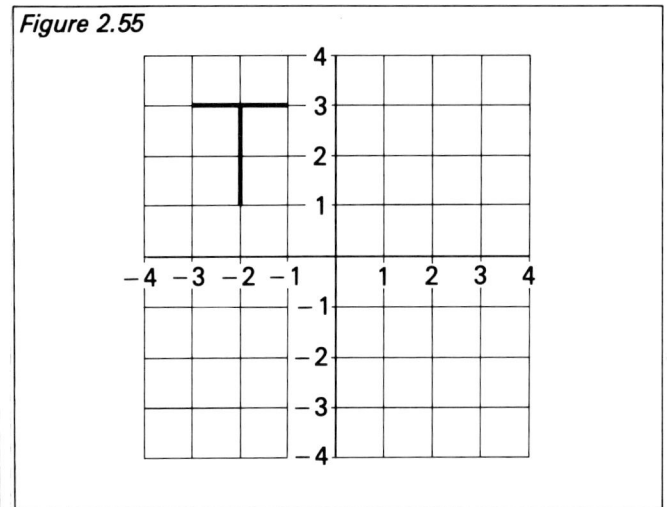

Figure 2.55

2 Rotate the triangle in Figure 2.56 by 90° in an anti-clockwise direction followed by a reflection in the x-axis. Describe a single transformation that can replace these two.

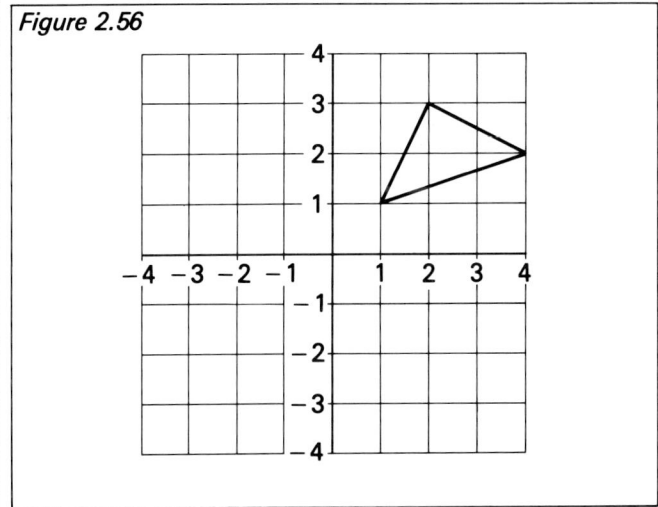

Figure 2.56

3 In Figure 2.57 what single transformation could replace the two transformations when the shape given is first rotated by 180° about the origin followed by a reflection in the x-axis?

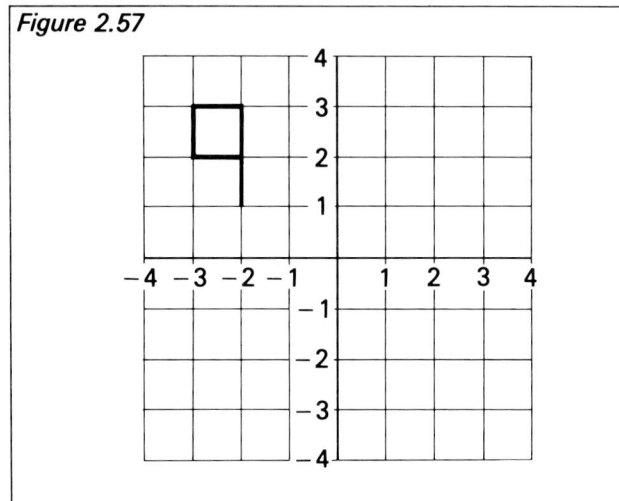

Figure 2.57

4 Copy the letter K in Figure 2.58. Draw in the reflection in the y-axis followed by a translation of $\begin{pmatrix} 0 \\ -4 \end{pmatrix}$. What single transformation could replace the two stated?

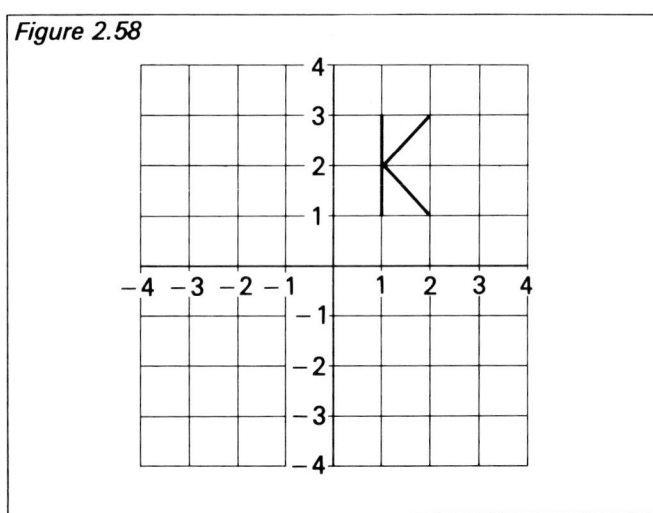

Figure 2.58

5 Reflect the square shown in Figure 2.59 in the mirror line $y = x$ then apply a 90° clockwise rotation about the origin. Is it possible to combine these two transformations into a single one and if so what would it be?

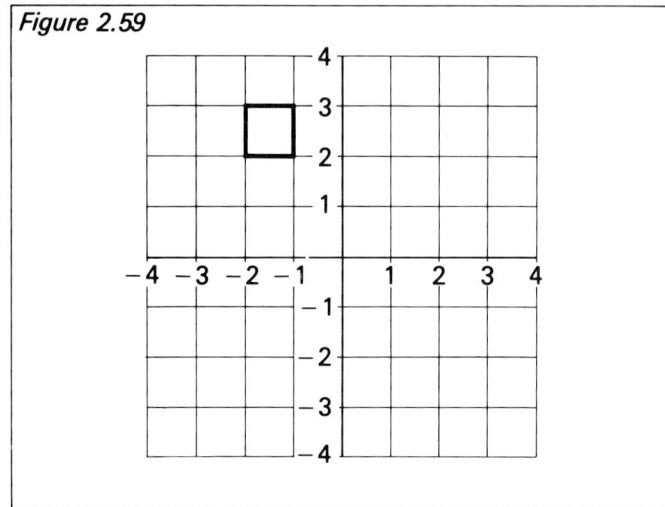

Figure 2.59

6 Apply the matrix $\begin{pmatrix} 0 & 2 \\ -2 & 0 \end{pmatrix}$ to the triangle ABC having coordinates A = (−2, 1), B = (−1, 2) and C = (−1, −1) (see Figure 2.60). Redraw the transformation stating what two separate transformations could have replaced it.

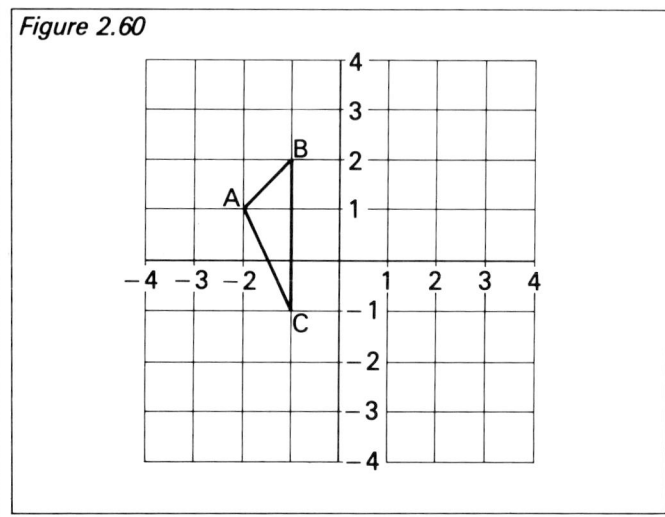

Figure 2.60

7 The quadrilateral KLMN has coordinates K = (1, −2), L = (3, −1), M = (4, −4) and N = (1, −4). Draw this figure and then apply the matrix $\begin{pmatrix} 0 & -\frac{1}{2} \\ -\frac{1}{2} & 0 \end{pmatrix}$ to each coordinate pair in turn. Draw the transformation which takes place and then state what two individual transformations would have given the same result.

8 A triangle RST has coordinates R = (−1, 3), S = (−2, 4) and T = (−4, 1). Apply the matrix $\begin{pmatrix} 1 & 0 \\ 0 & -1 \end{pmatrix}$ to the triangle and draw the transformation. Then apply matrix $\begin{pmatrix} -1 & 0 \\ 0 & -1 \end{pmatrix}$ to the new transformation redrawing a second transformation. What single transformation would replace the two given above?

9 Copy Figure 2.61 and then apply matrix $\begin{pmatrix} -1 & 0 \\ 0 & 1 \end{pmatrix}$. Redraw. Apply matrix $\begin{pmatrix} 1 & 0 \\ 0 & -1 \end{pmatrix}$ to the new transformation and again redraw. What single transformation could have been used instead of the two stated?

10 Copy Figure 2.62. Apply the matrix $\begin{pmatrix} 0 & -1 \\ -1 & 0 \end{pmatrix}$ and draw in the transformation. To this new transformation apply the second matrix $\begin{pmatrix} 0 & -1 \\ 1 & 0 \end{pmatrix}$. After redrawing state how the end result could have been obtained by a single transformation.

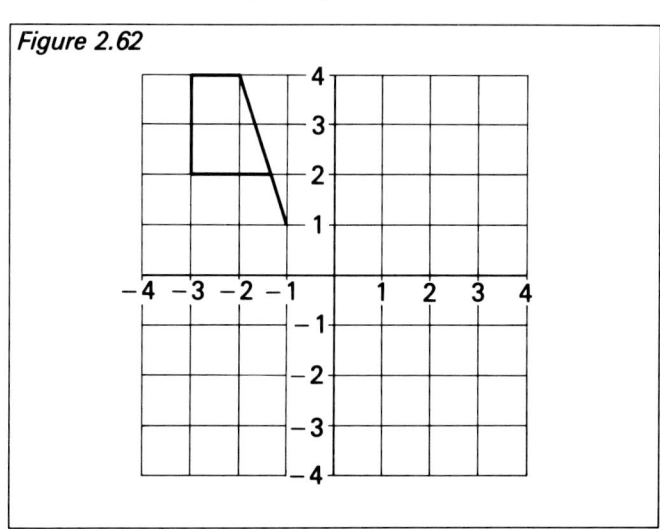

Figure 2.62

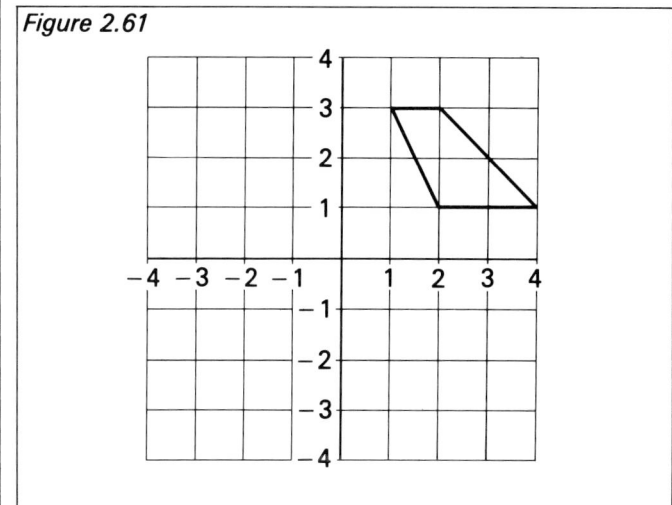

Figure 2.61

3 Vectors

General description

> *A VECTOR IS A QUANTITY HAVING BOTH MAGNITUDE AND DIRECTION. IT CAN BE SHOWN AS A LINE SEGMENT (SEE FIGURE 3.1).*

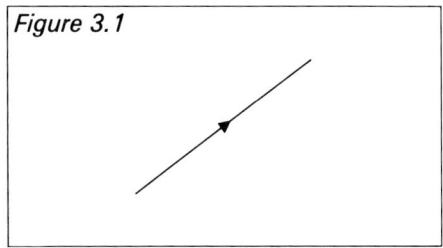

Figure 3.1

The magnitude of the vector is its *length*.
The direction is given by the *arrowhead* on the line segment.

In textbooks, vectors are usually denoted by *small* letters shown in bold type, **a**. But when you are working in your own books, you can use the equally acceptable description of a vector by writing a small letter underlined thus: a̲ or a (meaning vector a).

Vectors which have the same magnitude and direction are said to be equal, irrespective of their position in relation to one another. Vectors which have the same direction are *parallel* to one another (see Figure 3.2).

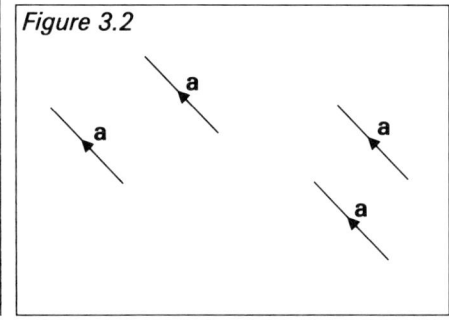

Figure 3.2

A vector can also be described another way.

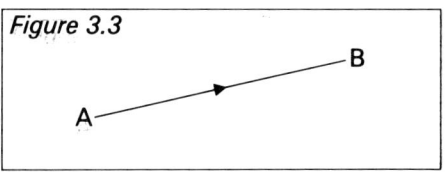

Figure 3.3

$\overrightarrow{AB}$ means a vector whose length is described by the line segment AB and whose direction is indicated by the arrow on top (i.e. from A to B). (Figure 3.3.)

Negative vectors

Let **x** represent the vector shown on the left of Figure 3.4. The negative of **x**, written −**x**, is a vector having the same magnitude but *the opposite direction* as on the right of Figure 3.4.

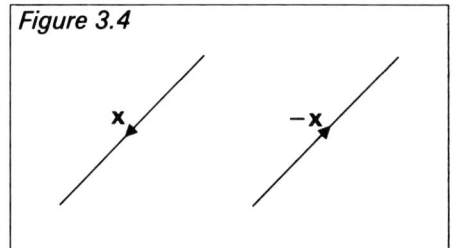

Figure 3.4

Scalars

Vectors which have the same direction and are therefore parallel may differ from one another in their length.

> *IF THE LENGTH OF ONE VECTOR IS KNOWN, THE LENGTH OF THE SECOND VECTOR IS RELATED TO IT BY A GIVEN FACTOR. THIS FACTOR IS KNOWN AS A SCALAR (SEE FIGURE 3.5).*

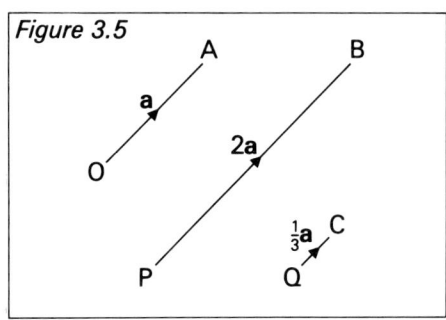

Figure 3.5

If $\overrightarrow{OA} = \mathbf{a}$, $\overrightarrow{PB} = 2\mathbf{a}$, $\overrightarrow{QC} = \frac{1}{3}\mathbf{a}$

$PB = 2OA$ and $QC = \frac{1}{3}OA$.

The values 2 and $\frac{1}{3}$ are scalar quantities and describe how the lengths of the two vectors relate to the length of the original vector.

Vectors and routes

Example 1

Look at Figure 3.6 and state how the route from O to C ($\overrightarrow{OC}$) can be described in terms of **a** and **b** given that $\overrightarrow{OA} = \mathbf{a}$ and $\overrightarrow{OB} = \mathbf{b}$.

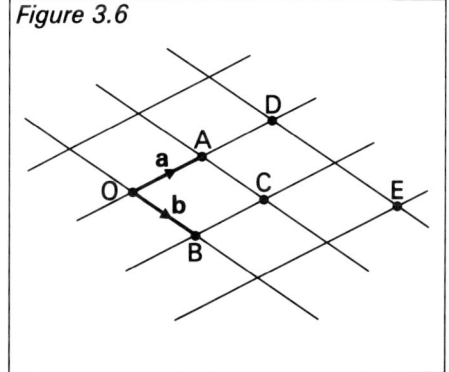

Figure 3.6

Solution

Assume that the only way from O to C is along the parallel lines shown

so $\overrightarrow{OC} = \overrightarrow{OA} + \overrightarrow{AC}$ *or* $\overrightarrow{OC} = \overrightarrow{OB} + \overrightarrow{BC}$

$\overrightarrow{OA} = \overrightarrow{BC} = \mathbf{a}$ (parallel and the same length)

and $\overrightarrow{OB} = \overrightarrow{AC} = \mathbf{b}$ (parallel and the same length).

Hence $\overrightarrow{OC} = \mathbf{a} + \mathbf{b}$ *or* $\overrightarrow{OC} = \mathbf{b} + \mathbf{a}$

It can be seen that the order in which the routes are taken does not affect the result.

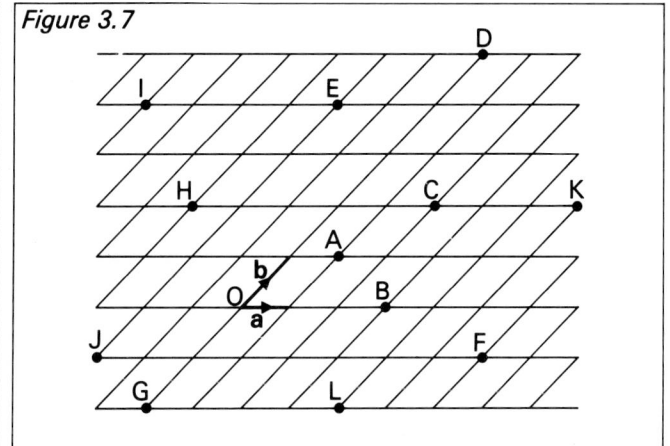

HENCE THE ADDITION OF VECTORS IS COMMUTATIVE.

Similarly $\overrightarrow{OD} = 2\overrightarrow{OA} = 2\mathbf{a}$
$\overrightarrow{DE} = 2\overrightarrow{OB} = 2\mathbf{b}$

and $\overrightarrow{OE} = \overrightarrow{OD} + \overrightarrow{DE}$
$= 2\mathbf{a} + 2\mathbf{b}$

Exercise 3.1

1 Describe the following vectors in terms of **a** and **b** in Figure 3.7.
(a) $\overrightarrow{OA}$ (b) $\overrightarrow{OB}$ (c) $\overrightarrow{OC}$ (d) $\overrightarrow{OD}$
(e) $\overrightarrow{OE}$ (f) $\overrightarrow{OF}$ (g) $\overrightarrow{OG}$ (h) $\overrightarrow{OH}$
(i) $\overrightarrow{OI}$ (j) $\overrightarrow{OJ}$ (k) $\overrightarrow{OK}$ (l) $\overrightarrow{OL}$

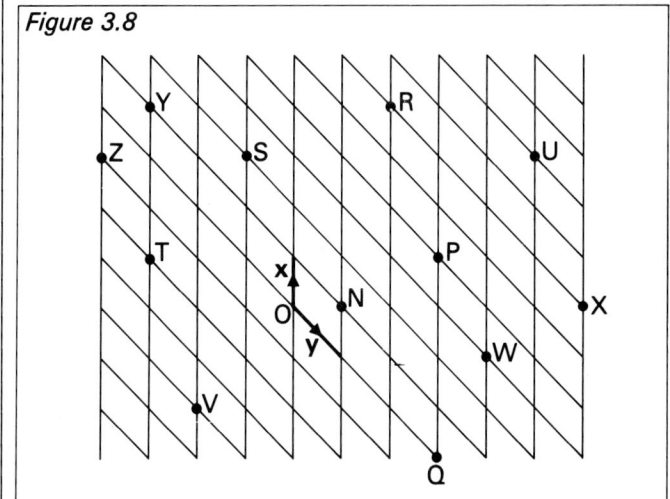

Figure 3.7

2 Describe the following vectors in terms of **x** and **y** in Figure 3.8.
(a) $\overrightarrow{ON}$ (b) $\overrightarrow{OP}$ (c) $\overrightarrow{OQ}$ (d) $\overrightarrow{OR}$
(e) $\overrightarrow{OS}$ (f) $\overrightarrow{OT}$ (g) $\overrightarrow{OU}$ (h) $\overrightarrow{OV}$
(i) $\overrightarrow{OW}$ (j) $\overrightarrow{OX}$ (k) $\overrightarrow{OY}$ (l) $\overrightarrow{OZ}$

Figure 3.8

Addition of vectors

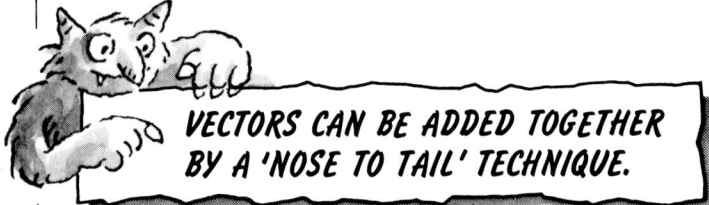

VECTORS CAN BE ADDED TOGETHER BY A 'NOSE TO TAIL' TECHNIQUE.

Example 2

$\overrightarrow{OA}$ and $\overrightarrow{XY}$ are two independent vectors (see Figure 3.9) and $\overrightarrow{OA} + \overrightarrow{XY} = \overrightarrow{OY}$. This addition is shown diagrammatically as follows.
(i) Draw a vector $\overrightarrow{OA}$. A is the 'nose' of $\overrightarrow{OA}$.
(ii) Draw in vector $\overrightarrow{XY}$ such that X, the 'tail' of $\overrightarrow{XY}$, is coincident with A.
(iii) Draw in the line $\overrightarrow{OY}$. This is known as the *resultant* and is indicated with a double arrowhead.

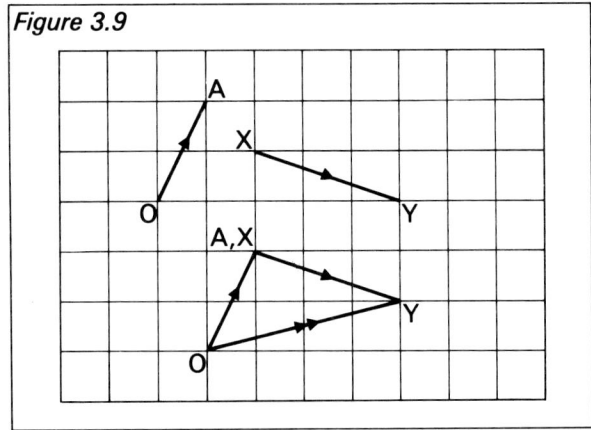

Figure 3.9

Example 3

Figure 3.10(a) shows two independent vectors **a** and **b**. Draw in the position of a point P such that $\overrightarrow{OP} = 2\textbf{a} + \textbf{b}$.

Solution (Figure 3.10(b))

OA is extended as far as X such that $\overrightarrow{OX} = 2\textbf{a}$. The 'tail' of $\overrightarrow{XP}$ which is a vector equal to $\overrightarrow{OB}$, is attached to the 'nose' of $\overrightarrow{OX}$. The position of point P is therefore fixed and the resultant $\overrightarrow{OP}$ can be drawn in.

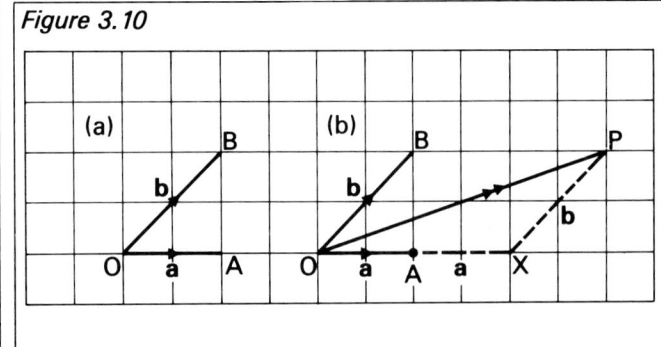

Figure 3.10

Example 4

Figure 3.11(a) shows two independent vectors **x** and **y**. Draw in the position of a point A such that $\overrightarrow{OA} = 2\textbf{y} - 2\textbf{x}$.

Solution (Figure 3.11(b))

OY is extended as far as K such that $\overrightarrow{OK} = 2\textbf{y}$. The 'tail' of vector $\overrightarrow{KA}$ is attached to the 'nose' of $\overrightarrow{OK}$. $\overrightarrow{KA}$ is twice the length of **x** and is in the opposite direction ($-2\textbf{x}$). The position of A is therefore fixed and the resultant $\overrightarrow{OA}$ can be drawn in.

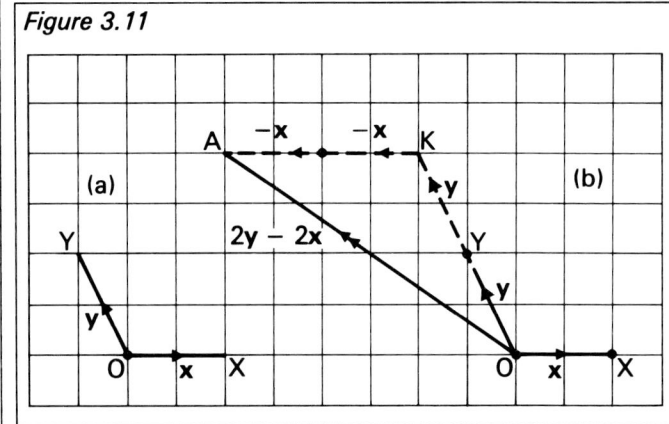

Figure 3.11

Exercise 3.2

Using the two independent vectors shown in Figure 3.12(a), draw on separate diagrams the following resultants.

1 $\overrightarrow{OA}$ such that $\overrightarrow{OA} = \textbf{a} + 2\textbf{b}$.

2 $\overrightarrow{OB}$ such that $\overrightarrow{OB} = 3\textbf{a} + \textbf{b}$.

3 $\overrightarrow{OC}$ such that $\overrightarrow{OC} = \textbf{a} - \textbf{b}$.

4 $\overrightarrow{OD}$ such that $\overrightarrow{OD} = \textbf{b} + 2\textbf{a}$.

5 $\overrightarrow{OE}$ such that $\overrightarrow{OE} = 2\textbf{b} - \textbf{a}$.

6 $\overrightarrow{OF}$ such that $\overrightarrow{OF} = 3\textbf{b} - 2\textbf{a}$.

7 $\overrightarrow{OG}$ such that $\overrightarrow{OG} = -\textbf{b} + 2\textbf{a}$.

8 $\overrightarrow{OH}$ such that $\overrightarrow{OH} = -\textbf{a} - \textbf{b}$.

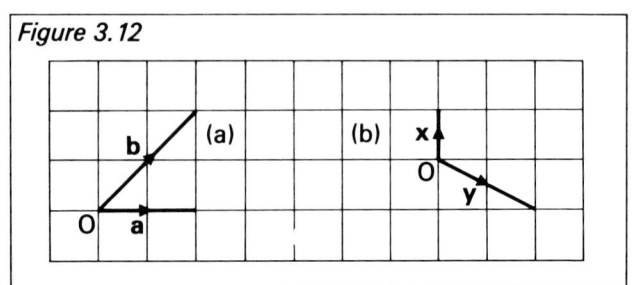

Figure 3.12

Similarly use Figure 3.12(b) to draw the following resultants.

9 $\overrightarrow{OT}$ such that $\overrightarrow{OT} = -2\mathbf{x} + 2\mathbf{y}$.

10 $\overrightarrow{OU}$ such that $\overrightarrow{OU} = 2\mathbf{x} - 4\mathbf{y}$.

11 $\overrightarrow{OV}$ such that $\overrightarrow{OV} = 3\mathbf{x} - \mathbf{y}$.

12 $\overrightarrow{OW}$ such that $\overrightarrow{OW} = 2\mathbf{y} - \mathbf{x}$.

13 $\overrightarrow{OX}$ such that $\overrightarrow{OX} = -\mathbf{x} - 2\mathbf{y}$.

14 $\overrightarrow{OY}$ such that $\overrightarrow{OY} = 4\mathbf{x} - 2\mathbf{y}$.

15 $\overrightarrow{OZ}$ such that $\overrightarrow{OZ} = 3\mathbf{y} + 2\mathbf{x}$.

Problems

Example 5

In Figure 3.13, $\overrightarrow{AB} = \mathbf{a}$, $\overrightarrow{AC} = \mathbf{b}$ and BD = DC.
Find the following in terms of $\mathbf{a}$ and $\mathbf{b}$ (a) $\overrightarrow{BA}$
(b) $\overrightarrow{BC}$ (c) $\overrightarrow{BD}$ (d) $\overrightarrow{CD}$ (e) $\overrightarrow{AD}$.

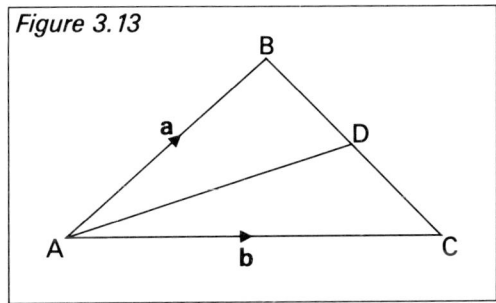

Figure 3.13

Solution

Express each question as one or more 'known vectors' before introducing $\mathbf{a}$ and $\mathbf{b}$.

(a) $\overrightarrow{BA} = -\overrightarrow{AB}$ ($\overrightarrow{AB}$ is a 'known vector')
      $= -\mathbf{a}$

(b) $\overrightarrow{BC} = \overrightarrow{BA} + \overrightarrow{AC}$ (both $\overrightarrow{BA}$ and $\overrightarrow{AC}$ are
                                            'known vectors')
      $= -\mathbf{a} + \mathbf{b}$
      $= \mathbf{b} - \mathbf{a}$

(c) $\overrightarrow{BD} = \frac{1}{2}\overrightarrow{BC}$ (since BD = DC is given and
                     $\overrightarrow{BC}$ is known from (b))
      $= \frac{1}{2}(\mathbf{b} - \mathbf{a})$
      $= \frac{1}{2}\mathbf{b} - \frac{1}{2}\mathbf{a}$

(d) $\overrightarrow{CD} = \frac{1}{2}\overrightarrow{CB}$
      $= -\frac{1}{2}\overrightarrow{BC}$ ($\overrightarrow{BC}$ is 'known vector' from (b))
      $= -\frac{1}{2}(\mathbf{b} - \mathbf{a})$
      $= -\frac{1}{2}\mathbf{b} + \frac{1}{2}\mathbf{a}$
      $= \frac{1}{2}\mathbf{a} - \frac{1}{2}\mathbf{b}$

(e) $\overrightarrow{AD} = \overrightarrow{AB} + \overrightarrow{BD}$ (both 'known vectors')
or $\overrightarrow{AD} = \overrightarrow{AC} + \overrightarrow{CD}$ (both 'known vectors')

$\overrightarrow{AD} = \overrightarrow{AB} + \overrightarrow{BD}$ or $\overrightarrow{AD} = \overrightarrow{AC} + \overrightarrow{CD}$
$= \mathbf{a} + \frac{1}{2}\mathbf{b} - \frac{1}{2}\mathbf{a}$ $= \mathbf{b} + \frac{1}{2}\mathbf{a} - \frac{1}{2}\mathbf{b}$
$= \frac{1}{2}\mathbf{a} + \frac{1}{2}\mathbf{b}$ $= \frac{1}{2}\mathbf{a} + \frac{1}{2}\mathbf{b}$

Exercise 3.3

1 In Figure 3.14, $\overrightarrow{OA} = \mathbf{a}$, $\overrightarrow{OC} = \mathbf{c}$ and AB = $\frac{2}{3}$AC.

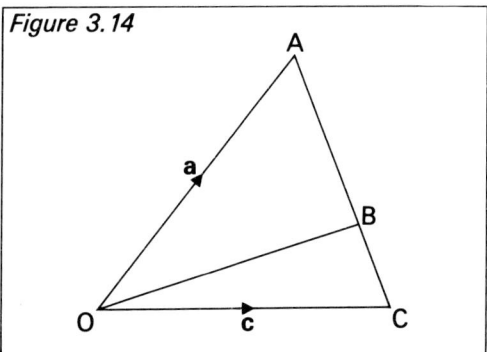

Figure 3.14

Find in terms of $\mathbf{a}$ and $\mathbf{c}$ (a) $\overrightarrow{AO}$ (b) $\overrightarrow{AC}$
(c) $\overrightarrow{AB}$ (d) $\overrightarrow{CB}$ (e) $\overrightarrow{OB}$.

2 In Figure 3.15, $\overrightarrow{OX} = \mathbf{x}$, $\overrightarrow{OY} = \mathbf{y}$ and OY = YZ.

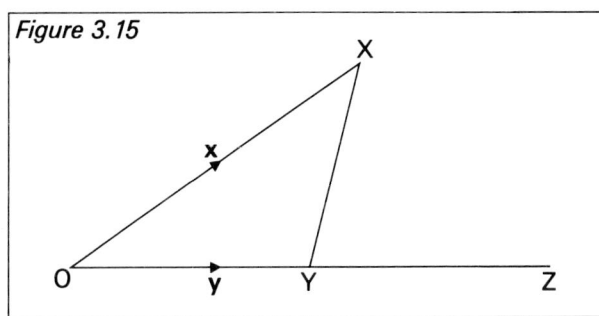

Figure 3.15

Express in terms of $\mathbf{x}$ and $\mathbf{y}$.   (a) $\overrightarrow{XO}$ (b) $\overrightarrow{XY}$
(c) $\overrightarrow{YZ}$ (d) $\overrightarrow{OZ}$ (e) $\overrightarrow{XZ}$.

3 In Figure 3.16, $\overrightarrow{AC} = \mathbf{a}$, $\overrightarrow{BC} = \mathbf{b}$ and CD = $\frac{1}{2}$AC.
E is the midpoint of AB.

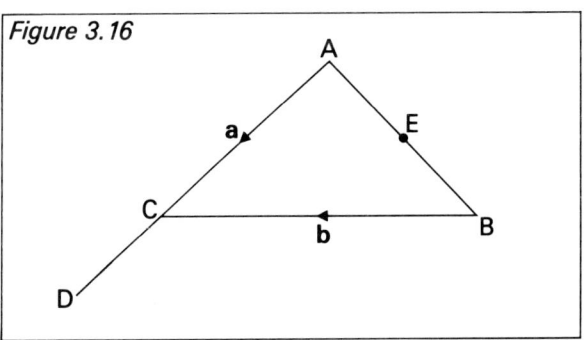

Figure 3.16

Express the following in terms of $\mathbf{a}$ and $\mathbf{b}$.
(a) $\overrightarrow{CB}$ (b) $\overrightarrow{AB}$ (c) $\overrightarrow{AD}$ (d) $\overrightarrow{AE}$ (e) $\overrightarrow{BE}$
(f) $\overrightarrow{DB}$ (g) $\overrightarrow{CE}$ (h) $\overrightarrow{ED}$.

4 Examine Figure 3.17 and the information given, then answer the questions, expressing the results in terms of **p** and **q**.
$\overrightarrow{OP} = \textbf{p}$, $\overrightarrow{OQ} = \textbf{q}$, $\overrightarrow{QR} = \frac{1}{2}\textbf{p}$, OR = RS.

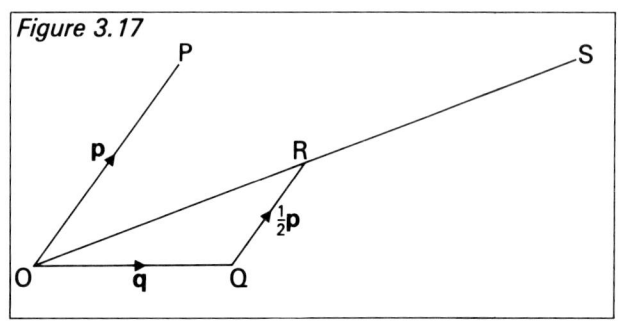

Figure 3.17

(a) $\overrightarrow{QO}$ (b) $\overrightarrow{QP}$ (c) $\overrightarrow{OR}$ (d) $\overrightarrow{OS}$ (e) $\overrightarrow{PR}$
(f) $\overrightarrow{QS}$ (g) $\overrightarrow{PS}$ (h) What can you say about OQ and PS?

5 In Figure 3.18, $\overrightarrow{OA} = \textbf{a}$, $\overrightarrow{OB} = \textbf{b}$, BD $= \frac{1}{2}$OB and BC $= \frac{1}{2}$AB.

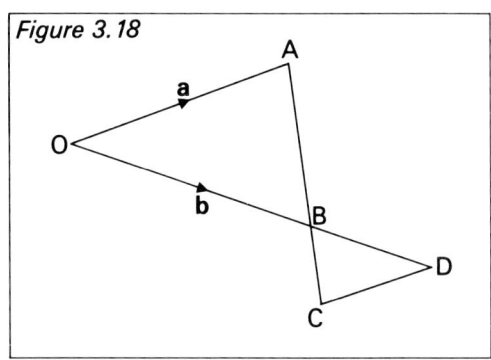

Figure 3.18

Find in terms of **a** and **b** the following (a) $\overrightarrow{AO}$
(b) $\overrightarrow{AB}$ (c) $\overrightarrow{BC}$ (d) $\overrightarrow{AC}$ (e) $\overrightarrow{BD}$ (f) $\overrightarrow{OD}$
(g) $\overrightarrow{AD}$ (h) $\overrightarrow{OC}$ (i) $\overrightarrow{CD}$ (j) What can be said about OA and CD?

6 Figure 3.19 shows a parallelogram PQRS. M and N are the midpoints of QR and RS respectively. $\overrightarrow{PQ} = \textbf{x}$ and $\overrightarrow{PS} = \textbf{y}$.

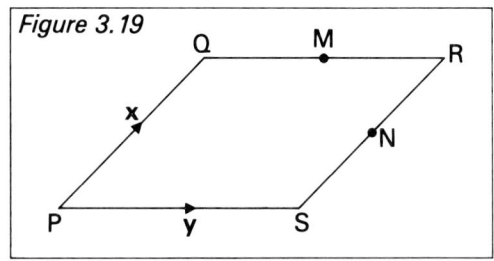

Figure 3.19

Express in terms of **x** and **y** (a) $\overrightarrow{QR}$ (b) $\overrightarrow{QP}$
(c) $\overrightarrow{RS}$ (d) $\overrightarrow{PR}$ (e) $\overrightarrow{RM}$ (f) $\overrightarrow{SN}$ (g) $\overrightarrow{PM}$ (h) $\overrightarrow{PN}$
(i) $\overrightarrow{QN}$ (j) $\overrightarrow{MS}$ (k) $\overrightarrow{MN}$.

7 In triangle OGH (Figure 3.20) M is the midpoint of OG, and GK $= \frac{3}{4}$GH. $\overrightarrow{OG} = \textbf{g}$, $\overrightarrow{OH} = \textbf{h}$.

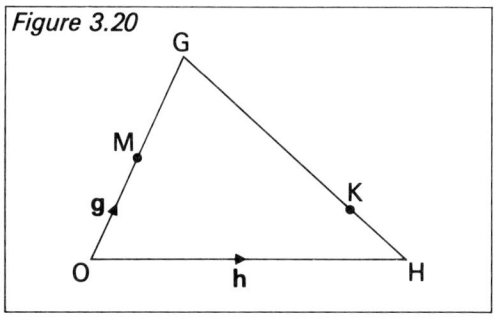

Figure 3.20

Find in terms of **g** and **h** (a) $\overrightarrow{GO}$ (b) $\overrightarrow{MO}$
(c) $\overrightarrow{GH}$ (d) $\overrightarrow{GK}$ (e) $\overrightarrow{HK}$ (f) $\overrightarrow{MH}$ (g) $\overrightarrow{KO}$
(h) $\overrightarrow{MK}$.

8 In Figure 3.21, F is the midpoint of DC. $\overrightarrow{OA} = \textbf{a}$, $\overrightarrow{BC} = 2\textbf{a}$, $\overrightarrow{AB} = \textbf{b}$, $\overrightarrow{OD} = \textbf{c}$.

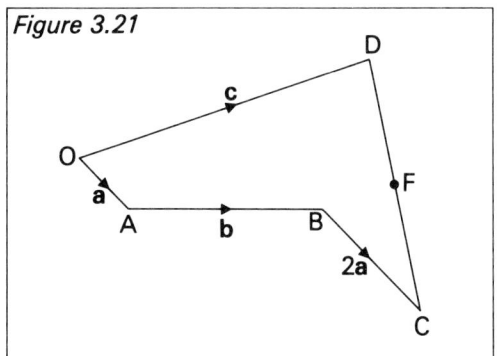

Figure 3.21

Express in terms of **a**, **b** and **c** (a) $\overrightarrow{DO}$ (b) $\overrightarrow{DA}$
(c) $\overrightarrow{OB}$ (d) $\overrightarrow{AC}$ (e) $\overrightarrow{DB}$ (f) $\overrightarrow{DC}$ (g) $\overrightarrow{DF}$ (h) $\overrightarrow{OF}$
(i) $\overrightarrow{FA}$ (j) $\overrightarrow{BF}$.

9 In Figure 3.22, $\overrightarrow{OA} = \textbf{a}$, $\overrightarrow{OB} = \textbf{b}$, OE $= \frac{1}{3}$OA, AD $= \frac{1}{3}$AB, OB = BC.

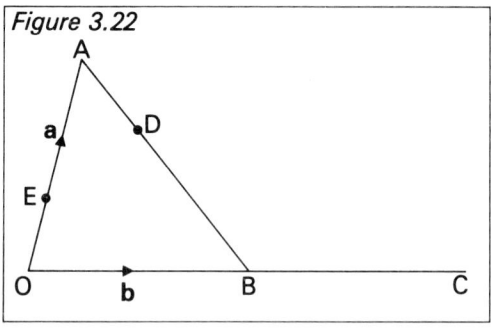

Figure 3.22

Express the following in terms of **a** and **b**.
(a) $\overrightarrow{AO}$ (b) $\overrightarrow{AE}$ (c) $\overrightarrow{AB}$ (d) $\overrightarrow{AD}$ (e) $\overrightarrow{BD}$
(f) $\overrightarrow{OC}$ (g) $\overrightarrow{CD}$ (h) $\overrightarrow{AC}$ (i) $\overrightarrow{EC}$ (j) $\overrightarrow{DE}$.

10 In Figure 3.23, M is the midpoint of XY and Y is the midpoint of OZ. $\overrightarrow{OY} = \textbf{y}$, $\overrightarrow{OX} = \textbf{x}$, XN $= \frac{1}{3}$XZ.

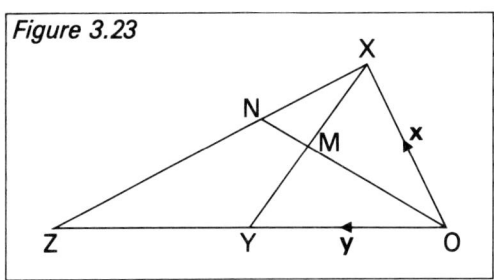

Figure 3.23

Using **x** and **y** only, find (a) $\overrightarrow{OZ}$ (b) $\overrightarrow{YO}$
(c) $\overrightarrow{YX}$ (d) $\overrightarrow{MX}$ (e) $\overrightarrow{MO}$ (f) $\overrightarrow{NZ}$ (g) $\overrightarrow{NO}$
(h) $\overrightarrow{NM}$ (i) $\overrightarrow{YN}$ (j) $\overrightarrow{ZM}$.

Vectors and coordinates

A VECTOR CAN BE EXPRESSED AS AN ORDERED PAIR OF NUMBERS IN THE FORM OF A <u>COLUMN MATRIX</u>, THUS: $\begin{pmatrix} x \\ y \end{pmatrix}$.
A VECTOR IN THIS FORM IS EASY TO REPRESENT GRAPHICALLY.

Example 6

Let $\mathbf{a} = \overrightarrow{XY} = \begin{pmatrix} 2 \\ 3 \end{pmatrix}$ (see Figure 3.24).

The position of X can be any point in the x–y plane. It can then be shown that 2 units from X parallel to the x-axis followed by 3 units parallel to the y-axis gives the vector $\begin{pmatrix} 2 \\ 3 \end{pmatrix}$. The position of Y is then located and a straight line joining X to Y, with an arrowhead to indicate the direction, completes the description of the vector.

Similarly, $\mathbf{b} = \overrightarrow{VW} = \begin{pmatrix} -3 \\ -1 \end{pmatrix}$, in Figure 3.24.

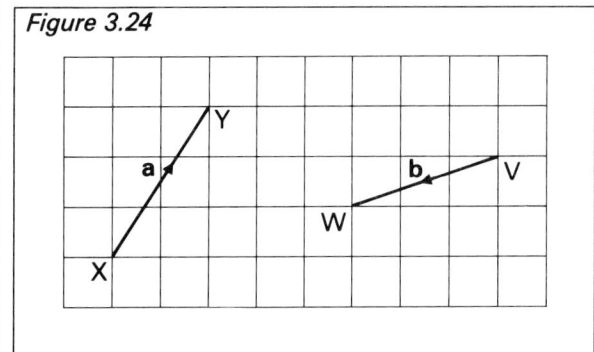

Figure 3.24

THE ADDITION OF THE TWO VECTORS IN FIGURE 3.24 CAN BE SHOWN DIAGRAMMATICALLY BY THE 'NOSE TO TAIL' TECHNIQUE (SEE FIGURE 3.25).

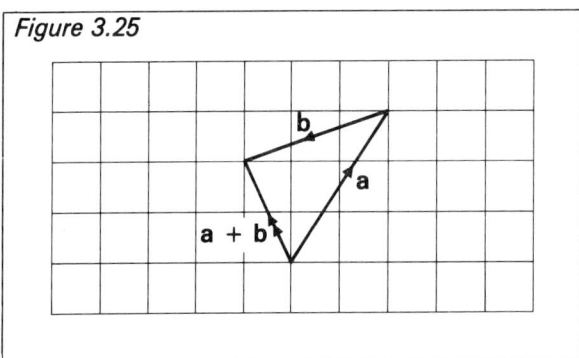

Figure 3.25

However the solution can be achieved non-graphically by adding together the column matrices representing **a** and **b**.

Thus $\mathbf{a} + \mathbf{b} = \begin{pmatrix} 2 \\ 3 \end{pmatrix} + \begin{pmatrix} -3 \\ -1 \end{pmatrix} = \begin{pmatrix} -1 \\ 2 \end{pmatrix}$

Verification can be obtained from Figure 3.25, since vector $(\mathbf{a} + \mathbf{b})$ is in fact one unit parallel to the x-axis in a negative direction followed by two units parallel to the y-axis in a positive direction.

Exercise 3.4

1 Express each vector in Figure 3.26 in the form $\binom{x}{y}$.

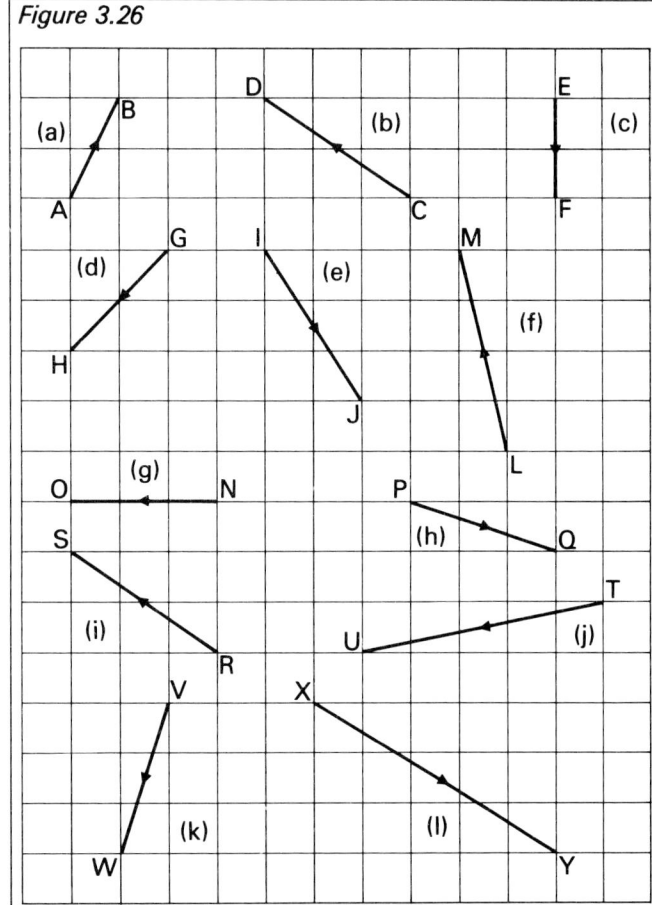

Figure 3.26

2 Use graph paper to represent the following vectors

(a) $\overrightarrow{AB} = \binom{5}{2}$ (b) $\overrightarrow{CD} = \binom{1}{4}$

(c) $\overrightarrow{EF} = \binom{-3}{2}$ (d) $\overrightarrow{GH} = \binom{0}{4}$

(e) $\overrightarrow{IJ} = \binom{-3}{-2}$ (f) $\overrightarrow{KL} = \binom{6}{-1}$

(g) $\overrightarrow{MN} = \binom{-5}{6}$ (h) $\overrightarrow{OP} = \binom{5}{0}$

(i) $\overrightarrow{QR} = \binom{3}{-1}$ (j) $\overrightarrow{ST} = \binom{-1}{-7}$

(k) $\overrightarrow{UV} = \binom{-7}{0}$ (l) $\overrightarrow{WX} = \binom{7}{-4}$

(m) $\overrightarrow{YZ} = \binom{0}{-3}$

3 Let $\mathbf{a} = \binom{1}{2}$, $\mathbf{b} = \binom{1}{4}$, $\mathbf{c} = \binom{-3}{2}$, $\mathbf{d} = \binom{4}{-1}$ and $\mathbf{e} = \binom{-3}{-2}$. For each of the parts below draw on graph paper the two independent vectors given (using the 'nose to tail' technique) and the resultant. Write the resultant in the form $\binom{x}{y}$. Then verify the correctness of each resultant by adding together the column vectors representing each independent vector.

(a) $\mathbf{a}$, $\mathbf{b}$, $\mathbf{a+b}$ (b) $\mathbf{c}$, $\mathbf{d}$, $\mathbf{c+d}$ (c) $\mathbf{a}$, $\mathbf{e}$, $\mathbf{a+e}$
(d) $\mathbf{b}$, $\mathbf{c}$, $\mathbf{b+c}$ (e) $\mathbf{d}$, $\mathbf{e}$, $\mathbf{d+e}$ (f) $\mathbf{b}$, $\mathbf{d}$, $\mathbf{b+d}$
(g) $\mathbf{a}$, $\mathbf{c}$, $\mathbf{a+c}$ (h) $\mathbf{b}$, $\mathbf{e}$, $\mathbf{b+e}$ (i) $\mathbf{a}$, $\mathbf{d}$, $\mathbf{a+d}$
(j) $\mathbf{c}$, $\mathbf{e}$, $\mathbf{c+e}$

4 (a) Using the vectors in question 2, draw the following on graph paper   (i) $2\overrightarrow{AB}$ (ii) $3\overrightarrow{GH}$ (iii) $2\overrightarrow{CD}$ (iv) $1\frac{1}{2}\overrightarrow{KL}$ (v) $-\overrightarrow{ST}$ (vi) $4\overrightarrow{QR}$ (vii) $-\frac{1}{2}\overrightarrow{WX}$ (viii) $-3\overrightarrow{IJ}$ (ix) $2\frac{1}{2}\overrightarrow{AB}$ (x) $\frac{1}{3}\overrightarrow{YZ}$.

(b) If $\mathbf{a} = \binom{5}{2}$, $\mathbf{b} = \binom{0}{4}$, $\mathbf{c} = \binom{1}{4}$, $\mathbf{d} = \binom{6}{-1}$, $\mathbf{e} = \binom{-1}{-7}$, $\mathbf{f} = \binom{3}{-1}$, $\mathbf{g} = \binom{7}{-4}$, $\mathbf{h} = \binom{-3}{-2}$, $\mathbf{i} = \binom{0}{-3}$,

find the solution to each of the parts below by the addition of the matrices representing the individual vectors.

(i) $2\mathbf{a}+\mathbf{b}$ (ii) $2\mathbf{c}+\mathbf{d}$ (iii) $4\mathbf{f}-\mathbf{e}$
(iv) $\mathbf{g}+\frac{1}{3}\mathbf{i}$ (v) $\mathbf{f}-3\mathbf{h}$ (vi) $1\frac{1}{2}\mathbf{d}+2\frac{1}{2}\mathbf{a}$
(vii) $-\frac{1}{2}\mathbf{g}-3\mathbf{h}$ (viii) $3\mathbf{b}+\mathbf{e}$ (ix) $\mathbf{i}+1\frac{1}{2}\mathbf{d}+\mathbf{e}$
(x) $\mathbf{f}+3\mathbf{b}-\mathbf{e}+2\mathbf{a}$.

Position vectors

IT HAS ALREADY BEEN SHOWN THAT A VECTOR CAN BE EXPRESSED AS AN ORDERED PAIR IN THE FORM OF A COLUMN MATRIX $\binom{x}{y}$. A <u>POSITION VECTOR</u> IS ALSO GIVEN BY AN ORDERED PAIR $\binom{x}{y}$, BUT IN RELATION TO THE ORIGIN.

Example 7

If O is the origin and A has coordinates (3, 4), then
vector $\overrightarrow{OA} = \begin{pmatrix} 3 \\ 4 \end{pmatrix}$, (see Figure 3.27).

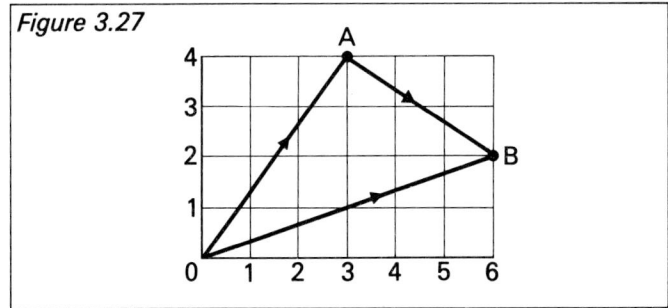

Figure 3.27

Similarly if $\overrightarrow{OB} = \begin{pmatrix} 6 \\ 2 \end{pmatrix}$, find $\overrightarrow{AB}$ in terms of $\begin{pmatrix} x \\ y \end{pmatrix}$.

Solution

$$\overrightarrow{AB} = \overrightarrow{AO} + \overrightarrow{OB}$$
$$= -\overrightarrow{OA} + \overrightarrow{OB}$$
$$= \overrightarrow{OB} - \overrightarrow{OA}$$
$$= \begin{pmatrix} 6 \\ 2 \end{pmatrix} - \begin{pmatrix} 3 \\ 4 \end{pmatrix}$$
$$\Rightarrow \overrightarrow{AB} = \begin{pmatrix} 3 \\ -2 \end{pmatrix}$$

Example 8

If $\overrightarrow{OP} = \begin{pmatrix} -3 \\ 8 \end{pmatrix}$ and $\overrightarrow{OQ} = \begin{pmatrix} -5 \\ -6 \end{pmatrix}$, find $\overrightarrow{PQ}$ in terms
of $\begin{pmatrix} x \\ y \end{pmatrix}$.

Solution

$$\overrightarrow{PQ} = \overrightarrow{PO} + \overrightarrow{OQ}$$
$$= -\overrightarrow{OP} + \overrightarrow{OQ}$$
$$= \overrightarrow{OQ} - \overrightarrow{OP}$$
$$= \begin{pmatrix} -5 \\ -6 \end{pmatrix} - \begin{pmatrix} -3 \\ 8 \end{pmatrix}$$
$$\overrightarrow{PQ} = \begin{pmatrix} -2 \\ -14 \end{pmatrix}$$

Exercise 3.5

1 If X is the point (4, 8) and Y is (9, 7), find the vector $\overrightarrow{XY}$.

2 If P is the point (−5, 3) and Q is (6, 4), find the vector $\overrightarrow{QP}$.

3 If C is the point (−3, 4) and D is (4, −6), find the vector $\overrightarrow{CD}$.

4 If K is the point (4, 9), L is (−3, −7) and M is (−2, 5), find the vectors $\overrightarrow{KL}$, $\overrightarrow{ML}$ and $\overrightarrow{KM}$.

5 Let X be the point (−2, −7), Y is (6, −3) and Z is (3, 8). Find the vectors $\overrightarrow{ZX}$, $\overrightarrow{YZ}$ and $\overrightarrow{YX}$.

6 S is the point (2, 3) and R is (6, 7). Find $\overrightarrow{SR}$ and $\overrightarrow{OX}$ such that X is the midpoint of SR.

7 If M is the point (−2, 4) and N is (−8, −8), find $\overrightarrow{NM}$. If X is the midpoint of MN, find $\overrightarrow{OX}$.

8 If A is the point (6, 7) and B the point (14, 17), find vector $\overrightarrow{OC}$ such that AC = CB.

9 Let V be the point (−3, −8) and W be the point (5, −4). Find the coordinates of Q, such that $\overrightarrow{OQ} = \overrightarrow{VW}$.

10 If P = (4, 3), Q = (10, −5) and R = (16, 19), find $\overrightarrow{PQ}$ and $\overrightarrow{QR}$. Find vectors $\overrightarrow{OM}$ and $\overrightarrow{ON}$ if M and N are the midpoints of PQ and QR respectively. What is the vector $\overrightarrow{MN}$?

4 Algebra 2

Solution of simultaneous equations

SIMULTANEOUS EQUATIONS ARE A PAIR OF EQUATIONS. THERE IS <u>ONE</u> UNIQUE SOLUTION (AN x AND A y VALUE) WHICH FITS <u>BOTH</u> EQUATIONS AT THE SAME TIME.

Example 1

Solve $x + 4y = 7$ (1)
        $x + 2y = 5$ (2)

Solution (elimination method)

x can be eliminated by subtracting equation (2) from equation (1).

Equation (1) $x + 4y = 7$
Equation (2) $\underline{x + 2y = 5-}$
              $2y = 2$
         $\Rightarrow y = 1$

Substitute $y = 1$ into either equation (1) or equation (2) to find x.

Equation (1) $x + 4y\ \ \ = 7$
              $x + 4 \times 1 = 7$
         $\Rightarrow x + 4\ \ \ \ = 7$
         $\Rightarrow x\ \ \ \ \ \ \ \ = 7 - 4$
         $\Rightarrow x\ \ \ \ \ \ \ \ = 3$

The ordered pair $(3, 1)$ is the solution to both equations.

Exercise 4.1

Solve the following pairs of simultaneous equations using the elimination method.

1 $x + 3y = 7$ **2** $5x + y = 11$
    $x + \ y = 3$ $3x + y = 7$

3 $x + 5y = 17$ **4** $x + y = 9$
    $x + 2y = 8$ $x - y = 1$

5 $2x + y = 5$ **6** $3x - 2y = -6$
    $2x - y = 3$ $x + 2y = 6$

7 $5x - 3y = 16$ **8** $5x - y = 14$
    $2x + 3y = -2$ $3x - y = 8$

9 $2x - 3y = 7$ **10** $x - 2y = 5$
    $4x - 3y = 11$ $5x - 2y = 9$

Example 2

Solve $4x - 2y = 14$ (1)
        $3x + 3y = 24$ (2)

Solution

If equation (1) is multiplied by 3 and equation (2) by 2, then the values of y will be the same (namely $6y$). The equations can then be added.

Equation (1) $4x - 2y = 14$
                              $3 \times$
                $\overline{12x - 6y = 42}$

Equation (2) $3x + 3y = 24$
                              $2 \times$
                $\overline{6x + 6y = 48}$

so $12x - 6y = 42$
                $\underline{6x + 6y = 48+}$
                $\overline{18x\ \ \ \ \ \ = 90}$
           $\Rightarrow\ x\ \ \ \ \ = 5$

Substitute $x = 5$ into equation (2)

           $3x + 3y = 24$
$\Rightarrow\ 3 \times 5 + 3y = 24$
   $\Rightarrow\ 15 + 3y = 24$
       $\Rightarrow\ 3y = 24 - 15$
       $\Rightarrow\ 3y = 9$
        $\Rightarrow\ y = 3$

The solution is the ordered pair $(5, 3)$.

Exercise 4.2

Solve the following pairs of simultaneous equations.

1 $x + 2y = 4$ **2** $4x - 4y = 4$
    $2x + \ y = 5$ $2x + \ y = 5$

3 $2x + 4y = 6$ **4** $2x - 5y = 10$
    $3x - 2y = -7$ $x - \ y = 2$

5 $x + 5y = 9$ **6** $3x - 5y = 10$
    $3x - \ y = -5$ $x - 2y = 4$

7 $3x - 6y = 33$
$x - 3y = 16$

8 $5x - 3y = 16$
$4x + 2y = 4$

9 $3x + 4y = -1$
$2x + 3y = -1$

10 $5x + 2y = 11$
$3x + 7y = -5$

Example 3

Solve $\quad 5x - 2y = 4 \qquad (1)$
$\qquad\quad 3x + 2y = 12 \qquad (2)$

Solution (inverse matrix method)

The two simultaneous equations can be rewritten as matrices in the following form

$$\begin{pmatrix} 5 & -2 \\ 3 & 2 \end{pmatrix}\begin{pmatrix} x \\ y \end{pmatrix} = \begin{pmatrix} 4 \\ 12 \end{pmatrix}$$

Find the inverse of matrix $\begin{pmatrix} 5 & -2 \\ 3 & 2 \end{pmatrix}$

(see Book 1, chapter 12).

If $\mathbf{M} = \begin{pmatrix} 5 & -2 \\ 3 & 2 \end{pmatrix}$,

$\mathbf{M}^{-1} = \dfrac{1}{10-(-6)}\begin{pmatrix} 2 & 2 \\ -3 & 5 \end{pmatrix} = \dfrac{1}{16}\begin{pmatrix} 2 & 2 \\ -3 & 5 \end{pmatrix}.$

MULTIPLY BOTH SIDES OF THE EQUATION BY THE INVERSE MATRIX.

$\dfrac{1}{16}\begin{pmatrix} 2 & 2 \\ -3 & 5 \end{pmatrix}\begin{pmatrix} 5 & -2 \\ 3 & 2 \end{pmatrix}\begin{pmatrix} x \\ y \end{pmatrix} = \dfrac{1}{16}\begin{pmatrix} 2 & 2 \\ -3 & 5 \end{pmatrix}\begin{pmatrix} 4 \\ 12 \end{pmatrix}$

But any matrix multiplied by its inverse gives the identity matrix for 2×2 matrices. This is $\begin{pmatrix} 1 & 0 \\ 0 & 1 \end{pmatrix}$.

Hence $\quad \begin{pmatrix} 1 & 0 \\ 0 & 1 \end{pmatrix}\begin{pmatrix} x \\ y \end{pmatrix} = \dfrac{1}{16}\begin{pmatrix} 2 & 2 \\ -3 & 5 \end{pmatrix}\begin{pmatrix} 4 \\ 12 \end{pmatrix}$

Multiplying the matrices on both sides gives

$$\begin{pmatrix} x \\ y \end{pmatrix} = \dfrac{1}{16}\begin{pmatrix} 8+24 \\ -12+60 \end{pmatrix}$$

$$\begin{pmatrix} x \\ y \end{pmatrix} = \dfrac{1}{16}\begin{pmatrix} 32 \\ 48 \end{pmatrix}$$

$$\begin{pmatrix} x \\ y \end{pmatrix} = \begin{pmatrix} 2 \\ 3 \end{pmatrix}$$

The solution is the ordered pair (2, 3).

Exercise 4.3

Solve the following pairs of simultaneous equations by the inverse matrix method.

1 $2x + y = 6$
$x + y = 3$

2 $x + y = 1$
$x - 2y = -5$

3 $x + 3y = -6$
$x - y = 2$

4 $3x + y = 5$
$2x - y = 0$

5 $2x + 5y = -6$
$2x - 5y = 14$

6 $2x - 3y = 1$
$x + y = 3$

7 $3x - 5y = -21$
$x + y = 1$

8 $5x - 4y = 18$
$2x + y = 2$

9 $7x - 3y = 16$
$x + y = -2$

10 $3x + 4y = 11$
$2x - y = 0$

11 $5x + 2y = 9$
$4x - y = 2$

12 $4x - y = -11$
$3x + 4y = 6$

Changing the subject of a formula

$y = 3x + 4 \qquad$ y is the subject.

$P = Q^2K \qquad$ P is the subject.

$V = \dfrac{(2x+y)^2}{3} \qquad$ V is the subject.

$F = \sqrt{\left(\dfrac{A}{B}\right)} \qquad$ F is the subject.

Example 4

Make x the subject of the formula $y = 3x + 4$.

Solution

$3x + 4 = y$

Subtract 4 from both sides

$3x + 4 - 4 = y - 4$
$\qquad\quad 3x = y - 4$

Divide both sides by 3

$\dfrac{\overset{1}{\cancel{3}}x}{\underset{1}{\cancel{3}}} = \dfrac{y-4}{3}$

$x = \dfrac{y-4}{3}$

Example 5

Make Q the subject of the formula $P = Q^2 K$.

Solution

$Q^2 K = P$

Divide both sides by K

$$\dfrac{Q^2 \overset{1}{\cancel{K}}}{\underset{1}{\cancel{K}}} = \dfrac{P}{K}$$

$$Q^2 = \dfrac{P}{K}$$

Square root both sides

$$Q = \sqrt{\left(\dfrac{P}{K}\right)}$$

Example 6

Make x the subject of the formula

$$V = \dfrac{(2x+y)^2}{3}.$$

Solution

$$V = \dfrac{(2x+y)^2}{3}$$

Multiply both sides by 3

$$3V = \dfrac{\overset{1}{\cancel{3}}(2x+y)^2}{\underset{1}{\cancel{3}}}$$

$$3V = (2x+y)^2$$

Square root both sides

$$\sqrt{(3V)} = 2x + y$$

but $2x + y = \sqrt{(3V)}$

Subtract y from both sides

$$2x + y - y = \sqrt{(3V)} - y$$
$$2x \quad = \sqrt{(3V)} - y$$

Divide both sides by 2

$$\dfrac{\overset{1}{\cancel{2}}x}{\underset{1}{\cancel{2}}} = \dfrac{\sqrt{(3V)} - y}{2}$$

$$x = \dfrac{\sqrt{(3V)} - y}{2}$$

Example 7

Make A the subject of the formula

$$F = \sqrt{\left(\dfrac{A}{B}\right)}.$$

Solution

$$\sqrt{\left(\dfrac{A}{B}\right)} = F$$

Square both sides

$$\dfrac{A}{B} = F^2$$

Multiply both sides by B

$$\overset{1}{\cancel{B}} \times \dfrac{A}{\underset{1}{\cancel{B}}} = B \times F^2$$

$$\Rightarrow \quad A = BF^2$$

Exercise 4.4

Make the letter in brackets the subject of each equation.

1 $a = 4b + 2$ (b) **2** $x = 5y - 3$ (y)

3 $g = \dfrac{3a}{5} - b$ (a) **4** $h = 3x^2$ (x)

5 $m = 2\sqrt{v}$ (v) **6** $p = (m+n)^2$ (n)

7 $a = bcd$ (c) **8** $k = 2(x^2 - y)$ (y)

9 $t = \sqrt{\left(\dfrac{a}{5}\right)}$ (a) **10** $l = \dfrac{PTR}{100}$ (R)

11 $F = \sqrt{(v+g)}$ (v) **12** $z = \dfrac{k+p}{4}$ (k)

13 $m = \dfrac{4u^2}{9}$ (u) **14** $t = \left(\dfrac{c+d}{3}\right)^2$ (d)

15 $V = \pi r^2 h$ (r) **16** $m = \sqrt{\left(\dfrac{5}{x}\right)}$ (x)

17 $k = \dfrac{1}{p^2}$ (p) **18** $e = 2\sqrt{(b-5)}$ (b)

19 $T = 2\pi \sqrt{\left(\dfrac{l}{g}\right)}$ (l) **20** $m = \dfrac{vtk^2}{4}$ (k)

Inequalities

THERE ARE TWO BASIC INEQUALITY SIGNS:
' > ' MEANS 'GREATER THAN'.
' < ' MEANS 'LESS THAN'.

THESE TWO SIGNS CAN ALSO BE COUPLED
WITH AN EQUAL SIGN, THUS:
' ⩾ ' MEANS 'GREATER THAN OR EQUAL TO'.
' ⩽ ' MEANS 'LESS THAN OR EQUAL TO'.

Since the statement, 'ten is greater than seven' is an equivalent statement to 'seven is less than ten', the following can be written

$$10 > 7 \quad \Leftrightarrow \quad 7 < 10$$

Example 8

Solve $2x + 3 > 9$.

Solution

$$2x + 3 - 3 > 9 - 3$$
$$\Rightarrow \quad 2x > 6$$
$$\Rightarrow \quad \frac{\overset{1}{\cancel{2}}x}{\underset{1}{\cancel{2}}} > \frac{6}{2}$$
$$\Rightarrow \quad x > 3$$

Example 9

Solve $3 \leqslant 4w - 9$.

Solution

$$3 + 9 \leqslant 4w - 9 + 9$$
$$\Rightarrow \quad 12 \leqslant 4w$$
$$\Rightarrow \quad \frac{12}{4} \leqslant \frac{\overset{1}{\cancel{4}}w}{\underset{1}{\cancel{4}}}$$
$$\Rightarrow \quad 3 \leqslant w$$
$$\Rightarrow \quad w \geqslant 3$$

Example 10

Solve $14 < 10 - 2x$.

Solution

$$14 - 10 < 10 - 2x - 10$$
$$\Rightarrow \quad 4 < -2x$$
$$\Rightarrow \quad \frac{4}{2} < \frac{-\overset{1}{\cancel{2}}x}{\underset{1}{\cancel{2}}}$$
$$\Rightarrow \quad 2 < -x$$
$$\Rightarrow \quad -x > 2$$

To find x we must multiply both sides of the inequality by -1.

$$\Rightarrow \quad x < -2$$

WHENEVER BOTH SIDES OF AN INEQUALITY
ARE MULTIPLIED BY −1, THE INEQUALITY
SIGN MUST BE CHANGED TO MAINTAIN
THE TRUTH OF THE STATEMENT.

Exercise 4.5

Solve the following inequalities.

1 $2x + 3 > 7$ **2** $3x + 4 \geqslant 13$

3 $5x - 3 > 12$ **4** $4 + 3x < 10$

5 $5 + 4x \leqslant 9$ **6** $5 < 2x - 1$

7 $-10 \geqslant 3x + 4$ **8** $-3 < 5 - 2x$

9 $2 \leqslant 4 - 3x$ **10** $7 \geqslant 5 + 2x$

11 $-5 < 2 + 3x$ **12** $4x + 10 < 2$

13 $5x + 4 \geqslant -6$ **14** $2x - 5 > -9$

15 $5 > 3 + 2x$ **16** $16 \leqslant 7 - 3x$

17 $10 \leqslant 5 - x$ **18** $22 > 2 - 4x$

19 $-7 > 3 + 2x$ **20** $-3 \leqslant 2x - 11$

5 Mappings, functions and graphs

Mappings

WHEN EACH MEMBER OF ONE SET CAN BE MATCHED WITH EXACTLY ONE MEMBER OF A SECOND SET, THEN THIS RELATIONSHIP CAN BE CALLED A __MAPPING__.

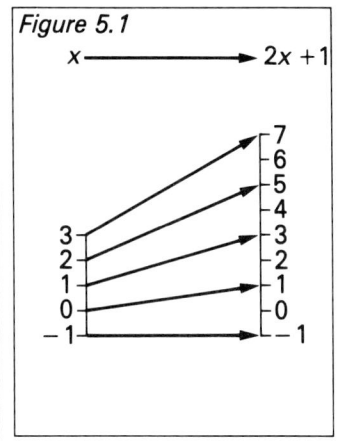

Example 1

Draw a mapping diagram for the mapping

$x \to 2x + 1$

from the set $\{-1, 0, 1, 2, 3\}$ to the set of real numbers.

Solution

$x \to 2x + 1$

This mapping means that for every value of x there is a corresponding value which is twice the value of x, plus one (see Figure 5.1).

Figure 5.1

Example 2

Draw the mapping diagram for

$x \to 2x^2 + 3$

from the set $\{-2, -1, 0, 1, 2\}$ to the set of real numbers.

Solution

This mapping means that for every value of x there is a corresponding value which is twice the value of x squared, plus three (see Figure 5.2).

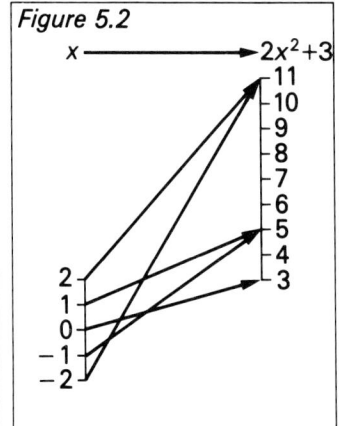

Figure 5.2

Exercise 5.1

In questions 1 to 10 draw the mapping diagram from the given set to the set of real numbers.

1 $x \to x - 3$ $\{2, 3, 4, 5, 6\}$

2 $x \to \dfrac{x}{2}$ $\{-4, -3, -2, -1, 0, 1, 2\}$

3 $x \to 3x + 1$ $\{-1, 0, 1, 2, 3\}$

4 $x \to 1 - x$ $\{-1, 0, 1, 2, 3\}$

5 $x \to 3x^2 - 2$ $\{-2, -1, 0, 1, 2\}$

6 $x \to \dfrac{3+x}{2}$ $\{-3, -2, -1, 0, 1, 2, 3\}$

7 $x \to 5 - 2x$ $\{-1, 0, 1, 2, 3, 4\}$

8 $x \to x^3$ $\{-2, -1, 0, 1, 2\}$

9 $x \to 2x - x^2$ $\{-2, -1, 0, 1, 2\}$

10 $x \to \dfrac{4x^2 - 3x}{2}$ $\{-2, -1, 0, 1, 2\}$

In questions 11 to 15 state the mapping which corresponds to the information on the mapping diagram.

11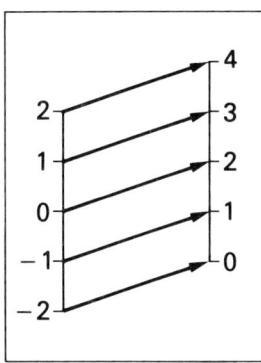

12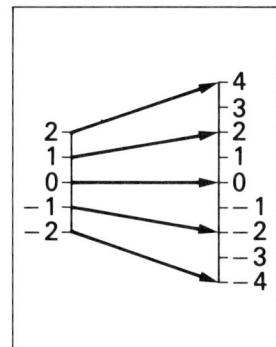

13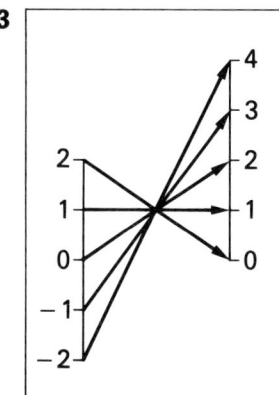

14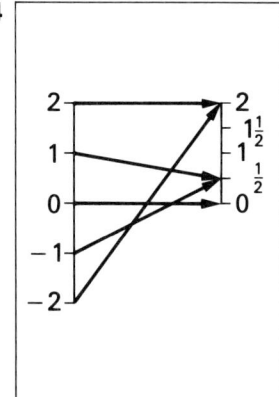

15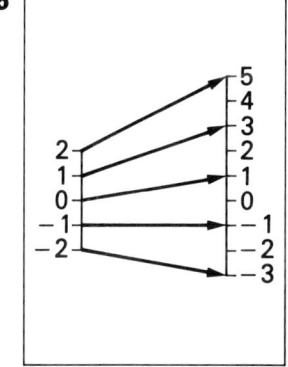

Functions

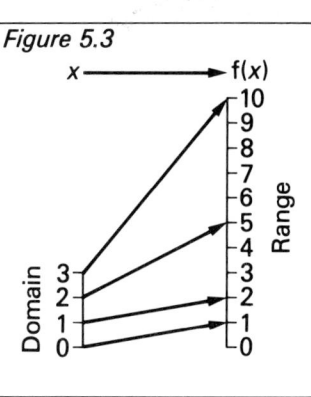

A MAPPING CAN ALSO BE REFERRED TO AS A FUNCTION, WHICH IS USUALLY DENOTED BY ONE OF THE LETTERS f, g OR h.

Example 3

Consider the statement

f: $x \rightarrow x^2 + 1$

This is read as 'f maps x onto $x^2 + 1$' and means that under the function f, any value x of a given set corresponds to the value of x squared plus one which belongs to a second set.

THE SET OF VALUES OF x WHICH IS USED IN THE MAPPING IS CALLED THE __DOMAIN__ AND THE CORRESPONDING VALUES UNDER THE FUNCTION ARE CALLED THE __IMAGES__. THE SET OF ALL THE IMAGES IS CALLED THE __RANGE__.

Consider the function f: $x \rightarrow x^2 + 1$ with a domain {0, 1, 2, 3}, then the image of each value within the domain can be evaluated more easily by writing the function as a formula.

Thus $f(x) = x^2 + 1$

$f(x)$ is read 'f of x' and is the value of any image under the function. So when $x = 0$ (replacing all x with 0)

$f(0) = 0^2 + 1 = 1$

and similarly when $x = 1$

$f(1) = 1^2 + 1 = 2$

when $x = 2$

$f(2) = 2^2 + 1 = 5$

when $x = 3$

$f(3) = 3^2 + 1 = 10$

Figure 5.3 shows the mapping diagram for this function and domain.

Figure 5.3

43

Example 4

Given the function $f(x) = x^2 + 1$, find f(4), f(−3) and the value of x when f(x) = 5.

Solution

f(4) means find the value of the function when $x = 4$.

$$f(4) = 4^2 + 1$$
$$= 16 + 1$$
$$= 17$$

and $f(-3) = (-3)^2 + 1$
$$= 9 + 1$$
$$= 10$$

When f(x) = 5, replace f(x) in the function $f(x) = x^2 + 1$ with 5.

$$5 = x^2 + 1$$
$$4 = x^2$$
$$x = \sqrt{4}$$
$$= \pm 2$$

Exercise 5.2

1 Find f(1) and f(−1) when $f(x) = 3x + 4$.

2 Find f(4) and f(−2) when $f(x) = 5 - 2x$.

3 If $g(x) = \dfrac{x^2 + 1}{2}$, calculate the values g(3) and g(−5).

4 Given $f(t) = t^3 + 1$, find f(1) and f(−2).

5 If $f(b) = 3(b - 2)$, find the values f(4) and f(−1).

6 Given $g(r) = 3r - 2$, find r when (a) g(r) = 4 and (b) g(r) = −11.

7 Find f(−2) and f(3) when $f(x) = (2x + 1)^2$.

8 Given $h(x) = \sqrt{(3x + 7)}$, find h(3) and h(14).

9 If $g(x) = x(x - 1)$, find the values of x when (a) g(x) = 0 and (b) g(x) = 2.

  (Hint: See factorisation of quadratics, Book 1, chapter 14.)

10 Given $h(x) = x^2 + 6x - 5$, find h(−3) and h(−6).

11 Given $g(b) = \dfrac{5b - 4}{3}$, calculate g(5) and the value of b when g(b) = 12.

12 Given $g(a) = a^2 - 2a + 3$, find g(−2) and g(3).

13 If $h(x) = x^2 - 3x$, find the values of h(3) and h(2). Calculate the values of x when h(x) = 0.

14 Find f(−2) and f(1) when $f(t) = t^2 + 4t - 5$. Find the values of t when f(t) = 0.

15 Given that $g(a) = 15 - 2a - a^2$, find g(3), g(4) and g(−1). Calculate the values of a when g(a) = 0.

Inverse functions

Consider the function $f: x \to 2x + 1$. When $x = 3$, the image under the function is found by replacing all values of x with 3, thus

$$f: 3 \to 2 \times 3 + 1$$
$$f: 3 \to 6 + 1$$
so $f: 3 \to 7$

Figure 5.4 shows this single mapping.

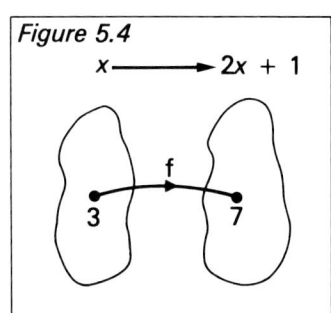

Figure 5.4

Sometimes it is useful to know what value of x generates a given image in the range. This can be resolved by finding the inverse function.

The function $f: x \to 2x + 1$ can be found by following the flow chart in Figure 5.5.

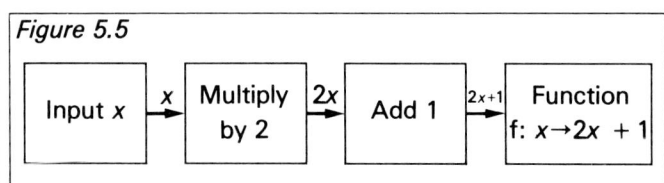

Figure 5.5

THE INVERSE FUNCTION IS FOUND BY REVERSING THE DIRECTION OF THE FLOW CHART AND USING THE OPPOSITE ARITHMETICAL OPERATIONS TO THOSE USED WHEN GENERATING THE FUNCTION (SEE FIGURE 5.6).

Figure 5.6

Under the function f: $x \to 2x + 1$ the image of 3 is 7. The inverse function should use the reverse path so that the image 7 is generated from a value in the domain of 3 (see Figure 5.7).

Thus $f^{-1}: x \to \dfrac{x-1}{2}$

$$f^{-1}: 7 \to \frac{7-1}{2}$$

$$f^{-1}: 7 \to \frac{6}{2}$$

$$f^{-1}: 7 \to 3$$

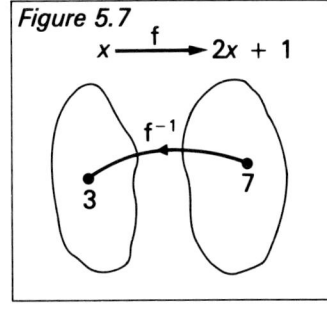

Figure 5.7

Exercise 5.3

Find the inverse functions of the following.

1 f: $x \to 3x$ **2** f: $x \to x - 5$

3 g: $x \to 8x$ **4** h: $x \to x + 6$

5 f: $x \to \dfrac{x}{5}$ **6** h: $x \to 2x + 4$

7 g: $x \to \dfrac{x+2}{5}$ **8** g: $x \to 3x - 5$

9 h: $x \to \dfrac{9}{x}$ **10** f: $x \to \dfrac{3x}{4}$

11 g: $x \to \dfrac{x}{3} + 1$ **12** h: $x \to \dfrac{5-x}{2}$

13 g: $x \to \dfrac{x-8}{3}$ **14** f: $x \to \dfrac{3x+1}{2}$

15 h: $x \to \dfrac{2x-4}{3}$ **16** f: $x \to \dfrac{3-2x}{4}$

Graphs of linear functions

Instead of writing a particular function as $f(x) = 2x - 1$, we often let $y = f(x)$, so the resulting equation becomes $y = 2x - 1$.

EQUATIONS OF THIS TYPE, WHERE THE HIGHEST POWER OF x IS A ONE, ARE CALLED <u>LINEAR</u>, WHICH MEANS THAT WHEN THEY ARE PLOTTED, A <u>STRAIGHT</u> LINE GRAPH IS OBTAINED.

When plotting a linear equation, it is usual to have at least four sets of ordered pairs (x, y).

Example 5

Plot the graph of the function $y = 2x - 1$ for $\{x: -2 \leqslant x \leqslant 5\}$ and use the graph to find the values of x when   (a) $y = -4$ and   (b) $y = 6$.

Solution

Any values of x from the domain between the limits -2 and 5 can be chosen.

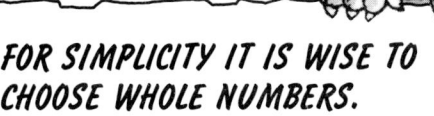

FOR SIMPLICITY IT IS WISE TO CHOOSE WHOLE NUMBERS.

Suppose $-2, 0, 2, 4, 5$ (5 values) are chosen, so when $x = -2$

$$y = 2(-2) - 1$$
$$= -4 - 1$$
$$= -5$$

The whole series of results is often presented in a table like the one below.

x	-2	0	2	4	5
$2x$	-4	0	4	8	10
-1	-1	-1	-1	-1	-1
y	-5	-1	3	7	9

The graph can now be plotted using the information from the table (see Figure 5.8). By following the dotted lines on the graph it can be seen that when $y = -4$, $x = -1\frac{1}{2}$ and when $y = 6$, $x = 3\frac{1}{2}$.

Figure 5.8

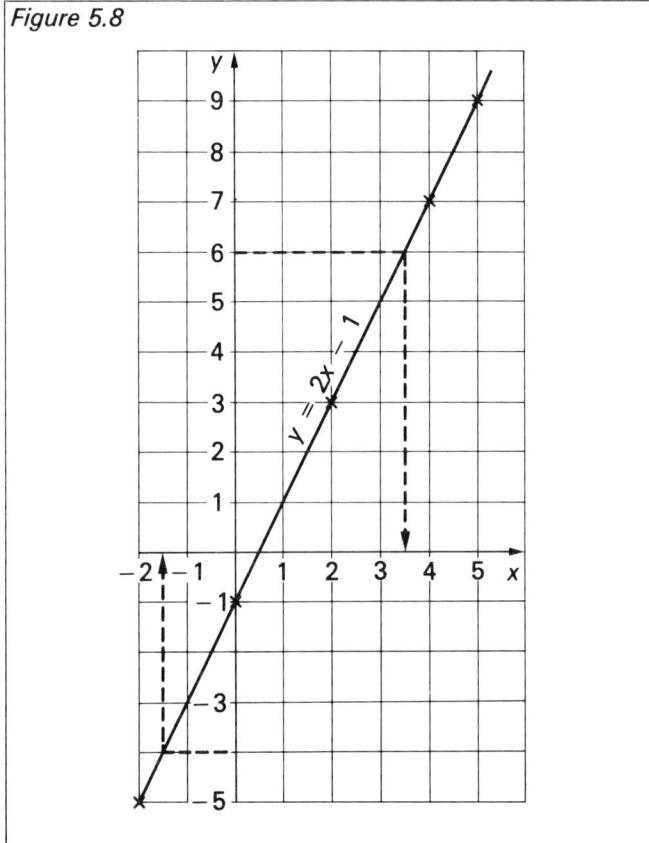

Exercise 5.4

1 Plot the function $y = 3x + 1$ for $\{x: -1 \leqslant x \leqslant 3\}$. Use the graph to find the value of x when $y = 6$.

2 Plot the graph $y = \frac{1}{2}x + 2$ for $\{x: -2 \leqslant x \leqslant 4\}$. Use the graph to evaluate y when $x = 2\frac{1}{2}$.

3 Plot the equation $y = 5 - 2x$ for $\{x: -1 \leqslant x \leqslant 3\}$. By using the graph find the value of x when $y = 4$.

4 Plot the graph of $y = 2 - \frac{1}{2}x$ for $\{x: -2 \leqslant x \leqslant 6\}$. Find from the graph the following values
(a) x when $y = 1\frac{1}{4}$ (b) y when $x = 2\frac{1}{2}$.

5 Draw the graph of $y = \frac{1}{4}x - 1$ for $\{x: -4 \leqslant x \leqslant 8\}$. Find the following values from the graph.
(a) x when $y = \frac{1}{2}$ (b) y when $x = 2$.

6 Draw the graph of $y = -2x - 3$ for $\{x: -3 \leqslant x \leqslant 3\}$. From the graph evaluate x when $y = -4$.

7 Plot the function $2y = x - 2$ for $\{x: -2 \leqslant x \leqslant 4\}$. Using the graph find x when $y = -1\frac{1}{4}$.

8 Draw the graph of $3y = 4 - x$ for $\{x: -2 \leqslant x \leqslant 7\}$. From the graph find the following values.
(a) x when $y = \frac{1}{2}$ (b) y when $x = 0$.

9 Plot the equation $x + 2y = 5$ for $\{x: -1 \leqslant x \leqslant 9\}$. Evaluate x when $y = 1\frac{1}{4}$ using the graph.

10 Plot the function $2x - 3y = 1$ for $\{x: -1 \leqslant x \leqslant 5\}$. Using the graph find the following values.
(a) x when $y = 1\frac{1}{2}$ (b) y when $x = 1\frac{1}{2}$.

11 Plot the two equations on the same graph
$$x + 4y = 7$$
$$\text{and} \quad x + 2y = 5$$
for $\{x: -1 \leqslant x \leqslant 5\}$. Write down the coordinates of the intersection of the two lines.

12 Plot the two equations on the same graph
$$3x - 2y = -6$$
$$\text{and} \quad x + 2y = 6$$
for $\{x: -3 \leqslant x \leqslant 3\}$. Which ordered pair satisfies both the given equations?

13 Draw both the equations given on the same graph and find the solution set of their intersection.
$$2x + 4y = 6$$
$$3x - 2y = -7$$
for $\{x: -3 \leqslant x \leqslant 4\}$.

14 Solve the pair of simultaneous equations by plotting each graph on the same axes.
$$5x - 3y = 16$$
$$4x + 2y = 4$$
for $\{x: -1 \leqslant x \leqslant 4\}$.

15 Solve the pair of simultaneous equations by plotting both graphs on the same axes.
$$x - 2y = 11$$
$$x - 3y = 16$$
for $\{x: -2 \leqslant x \leqslant 3\}$.

Gradient of a linear function

THE EQUATION OF A LINEAR FUNCTION HAS A GENERAL FORM, WHICH IS

$y = mx + c$

m is the gradient or steepness of the line.
c is called the intercept of the y-axis (the point where the line cuts the y-axis).

Examples 6

State the gradient and intercept of each of the following equations rearranging them first where necessary.

(a) $y = 2x + 4$.
Gradient $= 2$, intercept $= 4$.
(b) $y = \frac{1}{2}x - 3$.
Gradient $= \frac{1}{2}$, intercept $= -3$.
(c) $2y = 4x + 5$.
Rearrange so that the equation is in the form $y = mx + c$.

If $2y = 4x + 5$

$$\Rightarrow \quad y = \frac{4x + 5}{2}$$

$$\Rightarrow \quad y = 2x + 2\tfrac{1}{2}$$

Gradient $= 2$, intercept $= 2\frac{1}{2}$.

(d) $3x + 5y = 1$
Rearrange so that the equation is in the form $y = mx + c$.

If $3x + 5y = 1$

$$\Rightarrow \quad 5y = 1 - 3x$$

$$\Rightarrow \quad y = \frac{1 - 3x}{5}$$

$$\Rightarrow \quad y = \frac{1}{5} - \frac{3x}{5}$$

or $\quad y = -\frac{3}{5}x + \frac{1}{5}$

Gradient $= -\frac{3}{5}$, intercept $= \frac{1}{5}$.

THE GRADIENT AND INTERCEPT CAN ALSO BE FOUND BY PLOTTING THE GRAPH

Example 7

Plot the equation $y = 3x + 4$ for $\{x : -1 \leqslant x \leqslant 3\}$ and find the gradient and intercept.

Solution

See Figure 5.9.

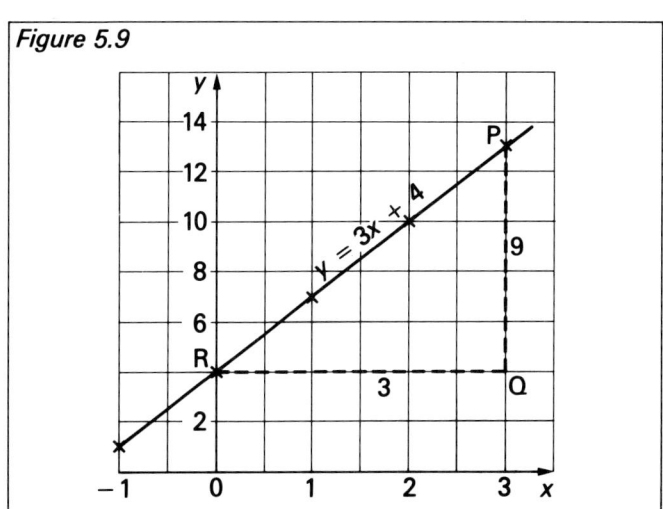

Figure 5.9

The gradient can be found by constructing a right-angled triangle, such as PQR, of a reasonable size and then calculating PQ and RQ.

The gradient $= \dfrac{PQ}{RQ} = \dfrac{9}{3} = 3$

The intercept, where the line passes through the y-axis, is seen to be 4.

Exercise 5.5

Write down the gradient and intercept of the equations in questions 1 to 15, rearranging them first into the form $y = mx + c$ where necessary.

1 $y = 5x - 1$ **2** $y = \frac{1}{2}x + 3$

3 $y = 4 + 2x$ **4** $y = 5 - 3x$

5 $-y = x + 2$ **6** $2y = x - 1$

7 $3y = 2x + 3$ **8** $-4y = 4x - 5$

9 $3y = 3 - 2x$ **10** $x + y = 4$

11 $2x + y = 3$ **12** $5x + 2y = 4$

13 $3x - 2y = 2$ **14** $-2x - 5y = 4$

15 $3x - 4y = -1$

In questions 16 to 20 (over the page) examine the graphs and write down the gradient, intercept and hence the equation of each line.

16

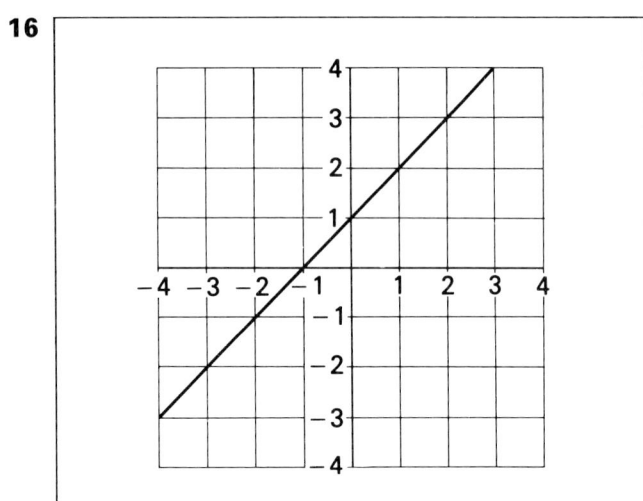

17

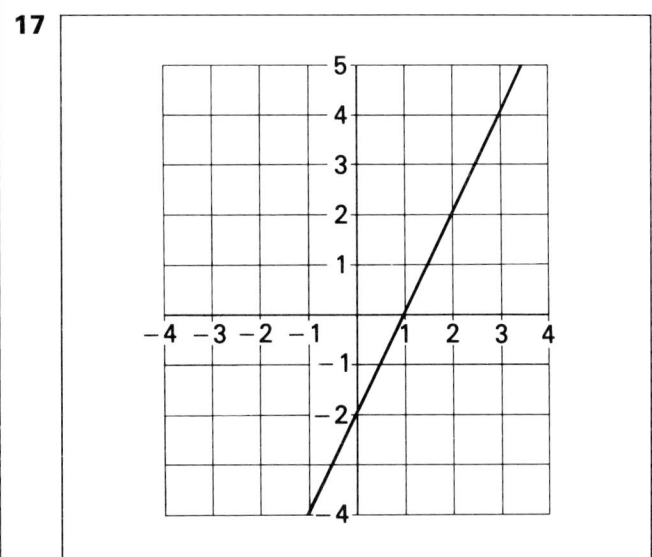

18

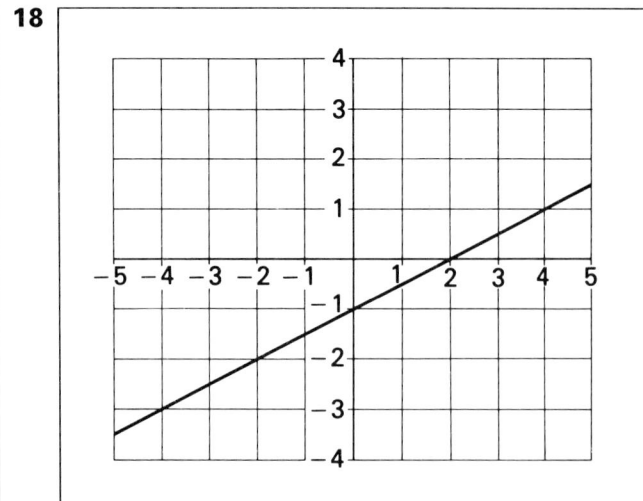

19

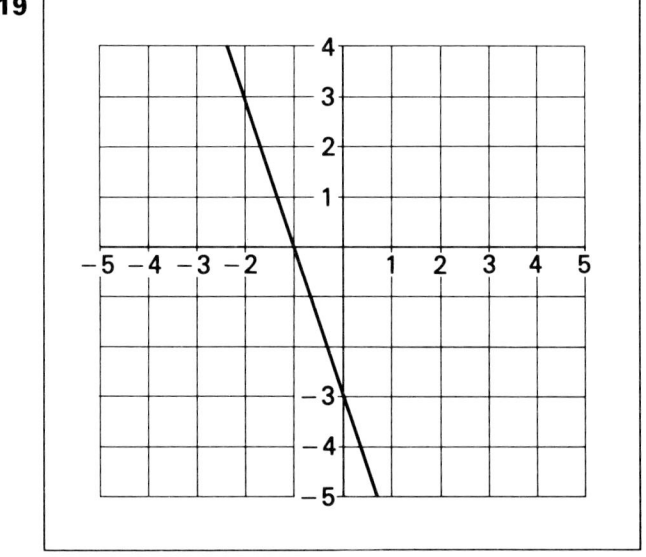

20

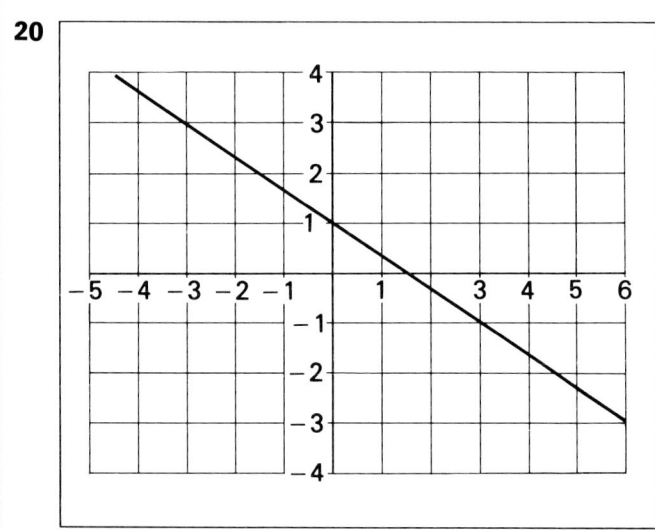

21 The table below gives values of the number of minutes it takes to cook a joint of beef whose weight is given in kilograms. By plotting the weight of beef along the x-axis and the time along the y-axis, complete the graph and find the gradient of the resulting line.

Time (min)	40	60	80	100	120	140	160
Weight (kg)	$\frac{1}{2}$	1	$1\frac{1}{2}$	2	$2\frac{1}{2}$	3	$3\frac{1}{2}$

22 Two quantities K and F are connected by a relationship. The table shows some of their corresponding values. Plot the relationship with K along the x-axis and F along the y-axis. Find the gradient of the resulting line.

K	4	6	10	12	20	26	28
F	15	22	36	43	71	92	99

23 The fixed cost (£P) of producing an article varies with the quantity produced (Q) according to the table below. Plot the graph with P along the y-axis and Q along the x-axis. Find the gradient and intercept of the line.

P (cost in £100's)	13	21	25	37	41	45
Q (quantity in 1000's)	2	4	5	8	9	10

24 A lorry travels a distance of s km in t hours according to the information given in the table. By first plotting s along the y-axis and t along the x-axis, find the gradient of the resulting line.

s	0	$22\frac{1}{2}$	$67\frac{1}{2}$	135	$247\frac{1}{2}$
t	0	$\frac{1}{2}$	$1\frac{1}{2}$	3	$5\frac{1}{2}$

Quadratics

A QUADRATIC FUNCTION IS ONE WHICH HAS THE GENERAL FORM

$$y = ax^2 + bx + c$$

WHERE a, b AND c ARE CONSTANTS. WHEN THE GRAPH IS DRAWN, IT IS A SMOOTH SYMMETRICAL CURVE CALLED A PARABOLA.

If the value of a (called the coefficient of x^2) is positive then the parabola generally looks like Figure 5.10.

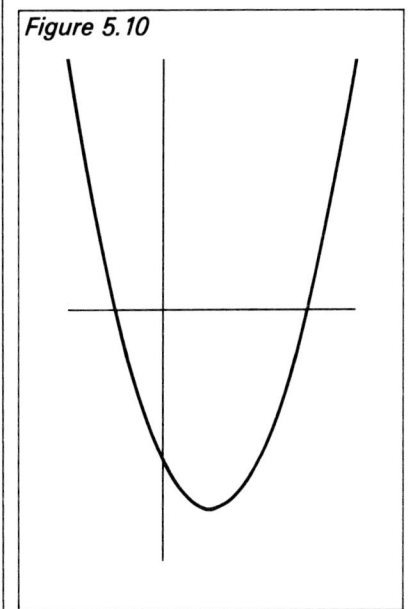
Figure 5.10

If the value of a is negative then the parabola is inverted and generally looks like Figure 5.11.

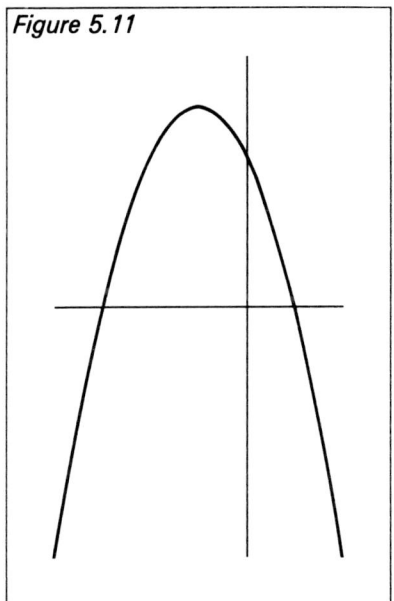
Figure 5.11

Example 8

Plot the curve $y = x^2 - x - 6$ for $\{x: -3 \leqslant x \leqslant 4\}$. Find the values of x when $y = 4$.

Solution

The equation can be sectionalised and tabulated. Corresponding values of x and y can be found by drawing and completing a table.

x	-3	-2	-1	0	1	2	3	4
x^2	9	4	1	0	1	4	9	16
$-x$	3	2	1	0	-1	-2	-3	-4
-6	-6	-6	-6	-6	-6	-6	-6	-6
y	6	0	-4	-6	-6	-4	0	6

Figure 5.12 shows the completed graph.

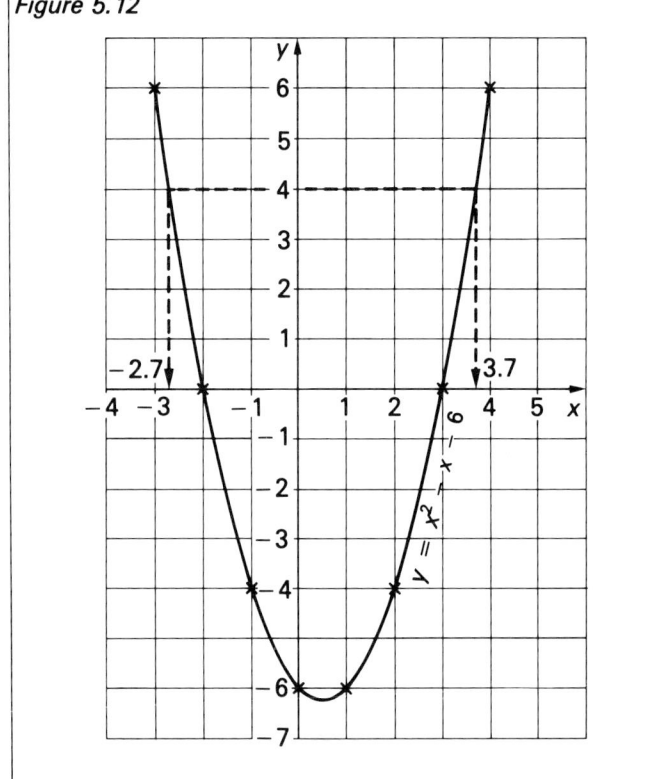

Figure 5.12

When $y = 4$, read off from the y-axis horizontally in both directions until the dotted line shown touches the curve. Then move vertically downwards and read off the values on the x-axis. Thus when $y = 4$, $x = -2.7$ and 3.7.

Example 9

A particle moves along a straight line from a point O according to the relationship $s = 2t^2 + 7t - 15$, where s is the distance in cm from O after a time of t seconds.
(a) Complete the table.

t	-7	-6	-5	-4	-3	-2	-1	0	1	2	3
$2t^2$	98			32			2			8	
$+7t$		-42		-28	-21			0			21
-15		-15		-15					-15	-15	
s				-11							

(b) Draw the graph of $s = 2t^2 + 7t - 15$ using a scale of 1 cm ≡ 1 unit on x-axis and 1 cm ≡ 5 units on y-axis.
(c) Use the graph to solve the equation $2t^2 + 7t - 15 = 0$.
(d) By drawing a tangent to the curve at the point (2, 7) and calculating its gradient, find the velocity of the particle at time $t = 2$ s.

Solution

(a) The table should be completed as below.

t	-7	-6	-5	-4	-3	-2	-1	0	1	2	3
$2t^2$	98	72	50	32	18	8	2	0	2	8	18
$+7t$	-49	-42	-35	-28	-21	-14	-7	0	7	14	21
-15	-15	-15	-15	-15	-15	-15	-15	-15	-15	-15	-15
s	34	15	0	-11	-18	-21	-20	-15	-6	7	24

(b) Figure 5.13 shows the completed graph.

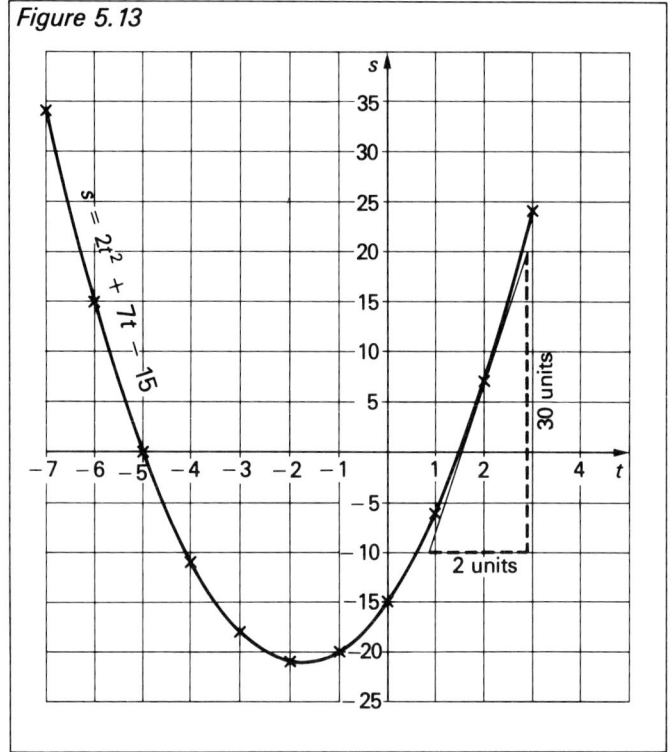

Figure 5.13

(c) The solution of $2t^2 + 7t - 15 = 0$ is where the curve $s = 2t^2 + 7t - 15$ and the line $s = 0$ intersect. There are two answers which have the values $t = -5$ and $t = 1\frac{1}{2}$.
(d) The gradient of the tangent at $t = 2$ is actually the velocity at $t = 2$.

Hence   velocity = gradient = $\dfrac{30}{2} = 15$ cm/s.

Exercise 5.6

1 Plot the graph of $y = x^2 - 9$ for values of $\{x: -4 \leqslant x \leqslant 4\}$. Find the values of x when $y = 3$ and $y = 0$.

2 Draw the graph of $y = 8 + 2x - x^2$ for the domain $\{x: -3 \leqslant x \leqslant 5\}$. Find the values of x when $y = -1$ and $y = 0$. What is the maximum value of the function?

3 Draw the graph of $y = 2x^2 - 7x - 4$ taking values of x between -2 and 6. Find the values of x when $y = 2$ and $y = 0$.

4 The velocity v in cm/s of a particle moving along a straight line after t seconds is given in the table below.

t	0	1	2	$2\frac{1}{2}$	3	4	5	6
v	4	0	-2	$-2\frac{1}{4}$	-2	0	4	10

(a) Using a scale of $2\,\text{cm} \equiv 1$ unit along the x-axis (t) and $1\,\text{cm} \equiv 1$ unit along the y-axis (v), plot the curve for the given table.
(b) Draw a tangent to the graph at the point where $t = 4$ and find its gradient. What is the name of the quantity you have just calculated?

5 Draw the graph of $y = 3x^2 + 5x - 2$ taking values of x between -4 and 2. Find the value of x when $y = 0$.

6 Plot the graph of $y = 4x^2 - 8x + 3$ for the domain $\{x: -1 \leqslant x \leqslant 4\}$. Hence find the solution to the equation $4x^2 - 8x + 3 = 0$.

7 Draw the graph of $y = -15 + 8x - x^2$ for the domain $\{x: 0 \leqslant x \leqslant 6\}$. Find the values of x when $y = 0$. What is the maximum value of the function?

8 The table below shows the distance travelled by a plane in the first 30 seconds of flight.

t(s)	0	5	10	15	20	25	30
d(m)	0	$12\frac{1}{2}$	50	$112\frac{1}{2}$	200	$312\frac{1}{2}$	450

(a) Using a scale of $2\,\text{cm} \equiv 5\,\text{s}$ horizontally and $1\,\text{cm} \equiv 25\,\text{m}$ vertically, plot the curve representing the information in the table.
(b) Draw a tangent to the curve at the point where $t = 20\,\text{s}$ and by calculating the gradient of this tangent, state the velocity of the plane when $t = 20\,\text{s}$.

9 The velocity (v) of a particle after a time t seconds is given by the relationship $v = 3t - t^2$. Copy and complete the table below and then

(a) draw the graph of velocity versus time for the values in the table using a scale of $2\,\text{cm} \equiv 1\,\text{s}$ on the x-axis and $1\,\text{cm} \equiv 1\,\text{cm/s}$ on the y-axis.
(b) What is the greatest velocity reached by the particle?
(c) At what time is the greatest velocity reached?
(d) By drawing a tangent and finding its gradient, calculate the acceleration of the particle after $3\,\text{s}$.

t(s)	0	1	2	3	4	5
$3t$		3				15
$-t^2$				-9		-25
v (cm/s)						-10

10 Draw a graph of $y = x^2 + 3x - 10$ for the domain $\{x: -6 \leqslant x \leqslant 3\}$ using a scale of $1\,\text{cm} \equiv 1$ unit on the x-axis and $1\,\text{cm} \equiv 2$ units on the y-axis.
   Now plot the equation $y = x - 7$ for the domain $\{x: -5 \leqslant x \leqslant 3\}$ on the same graph and hence find the solution of the equation $x^2 + 3x - 10 = x - 7$.

11 Plot the curve $y = 4 - x^2$ for the values of x between -3 and $+3$ using a scale of $2\,\text{cm} \equiv 1$ unit along the x-axis and $2\,\text{cm} \equiv 1$ unit along the y-axis.
(a) Draw in the line $y = 1$ and hence solve the equation $4 - x^2 = 1$.
(b) By counting squares or otherwise, find the approximate area bounded by the x-axis, the curve $y = 4 - x^2$ and the line $y = 1$.

12 Draw the graph of $y = (x - 2)^2$ by first completing the table below. Use a scale of $2\,\text{cm} \equiv 1$ unit in both the x and y directions.

x	0	$\frac{1}{2}$	1	$1\frac{1}{2}$	2	$2\frac{1}{2}$	3	$3\frac{1}{2}$	4
$x-2$	-2			$-\frac{1}{2}$			$1\frac{1}{2}$		
y	4			$\frac{1}{4}$			$2\frac{1}{4}$		

(a) Find the values of x when $y = 3$.
(b) By counting squares or otherwise, estimate the area bounded by the x-axis, the y-axis and the curve.

Distance–time graphs

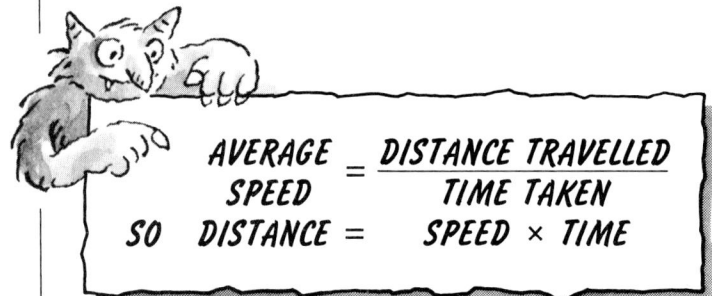

AVERAGE SPEED = DISTANCE TRAVELLED / TIME TAKEN
SO DISTANCE = SPEED × TIME

For constant speeds the distance travelled is always proportional to the time taken.

Example 10

A barge travels 25 km in 5 hours. Draw a distance–time graph to represent this information and find (a) the average speed of the barge (b) how long the barge takes to travel 16 km.

Solution

Assume that the barge travels equal distances each hour of its journey since no other information is available.

The graph representing this information is given in Figure 5.14.

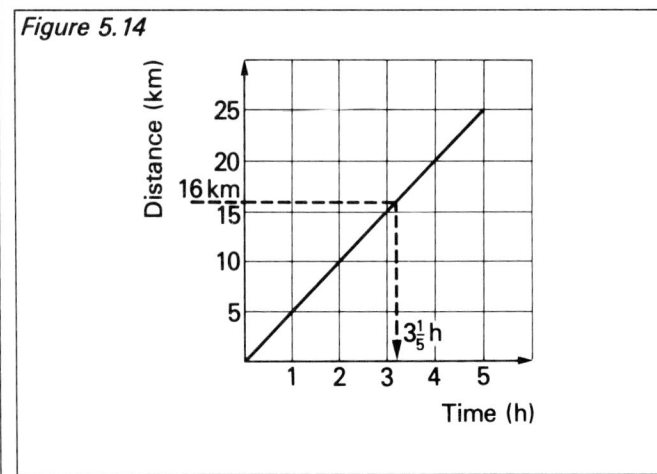

Figure 5.14

(a) Average speed = $\dfrac{\text{total distance travelled}}{\text{total time taken}}$

$= \dfrac{25}{5}$

$= 5 \text{ km/h}$.

(b) Read off from 16 km on the *y*-axis horizontally until the line of the graph is reached. Then drop a line vertically to touch the *x*-axis at $3\frac{1}{5}$ h. Hence the barge travels 16 km in 3 h 12 min.

Example 11

A car makes a journey of 300 km in 6 hours according to the graph in Figure 5.15.

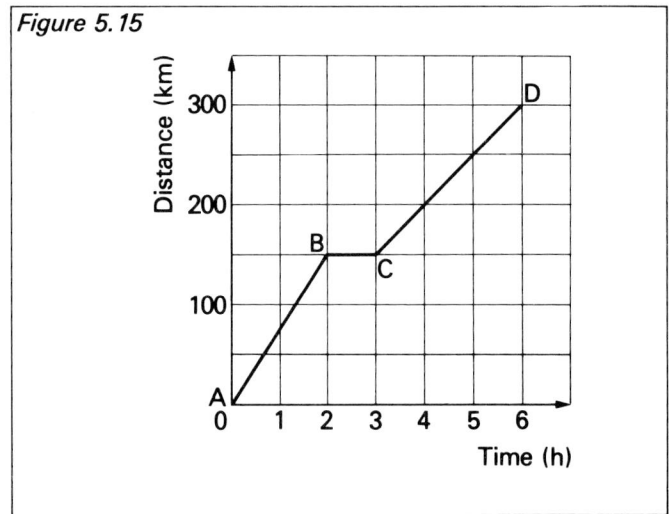

Figure 5.15

(a) What was the average speed for the section of journey A → B?
(b) Suggest what was happening during period B → C.
(c) What was the average speed for the section of journey C → D?
(d) What was the average speed for the whole journey?

Solution

Average speed = $\dfrac{\text{total distance travelled}}{\text{total time}}$

(a) Average speed A → B

$= \dfrac{150}{2} = 75 \text{ km/h}$.

(b) Car at rest for 1 h (perhaps a lunch stop).
(c) Average speed C → D

$= \dfrac{150}{3} = 50 \text{ km/h}$.

(d) Average speed A → D

$= \dfrac{300}{6} = 50 \text{ km/h}$.

Exercise 5.7

1 Jane cycles 21 km in 3 hours. She then rests for $\frac{1}{2}$ hour. She finally completes her journey by cycling a further 15 km in $2\frac{1}{2}$ hours.

 Plot this information on a graph and find
 (a) her average speed before she rested
 (b) her average speed after she rested
 (c) the average speed for the whole journey.

2 A man walks 15 km in 2 hours. He then waits $\frac{1}{4}$ hour for a bus before making the return journey in a further $\frac{1}{4}$ hour. Draw a graph of the journey and find   (a) the average speed while he was walking   (b) the average speed for the whole journey.

3 A private plane travels a distance of 1000 km in 5 hours. Plot a distance–time graph of the journey and find the average speed for the whole journey. Use your graph to find how long it took to cover the first 150 km.

4 A cruise liner covers 750 miles in 3 days. It then stays in port for 2 days before sailing a further 800 miles during the next 4 days. Another day's rest in port is completed before the liner starts out on its final section of the cruise which is 895 miles covered in a further 5 days.

   Draw a distance–time graph to show the progress of the cruise. Find the average speed of the three individual sections of the cruise and the average speed of the liner over the total voyage.

5 Figure 5.16 shows the progress of two trains. A goods train is shown by the line AE and a passenger train by the lines AB, BC, CD and DE. Find
   (a) how far apart the trains are at 11.00
   (b) the two times when the trains are the same distance from A
   (c) the average speed of the passenger train between 11.00 and 12.30.

6 A car travels 400 km in 6 hours. A motorbike sets off 2 hours later and completes the same journey in only 2 hours.

   How many km are they from the beginning of the journey when the bike overtakes the car?
   (Hint: draw the graph to show the two journeys together.)

7 Some hitch-hikers walk 10 km in 2 hours and then get a lift with a lorry and travel 135 km in 3 hours. Then they stop for some coffee in a motorway service station for 20 minutes before completing their journey of a further 95 km by sports car in 40 minutes.

   Draw a graph of the journey and find the following average speeds   (a) while walking
   (b) riding in the lorry   (c) riding in the sports car
   (d) total journey.

8 Figure 5.17 shows the progress of two hikers Smith and Jones. Smith follows the route A → F and Jones the route W → Z.
   (a) By how much longer does Smith rest than Jones?
   (b) What is Smith's average speed for the whole journey?
   (c) What is Jones's average speed for the whole journey?
   (d) When is the first time Jones overtakes Smith?
   (e) When is the first time Smith overtakes Jones?
   (f) How far are Smith and Jones apart at   (i) 7.30
   (ii) 8.45?
   (g) At what time does Jones overtake Smith for the second time?

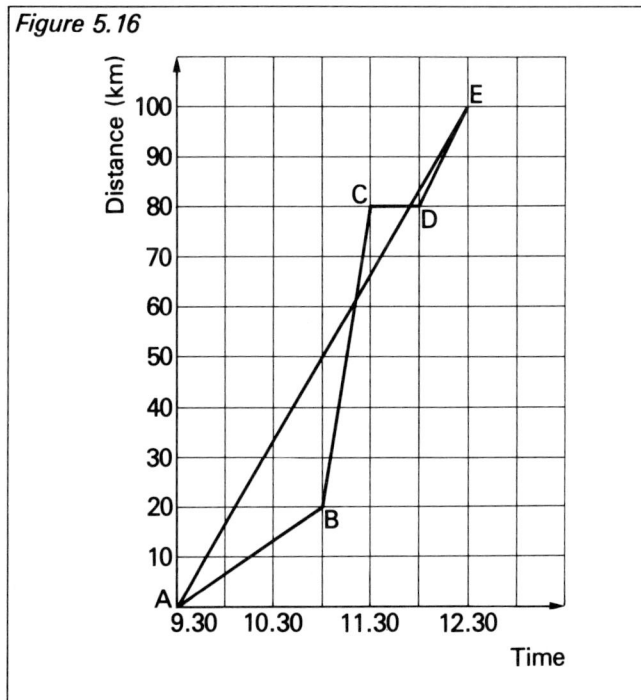

Figure 5.16

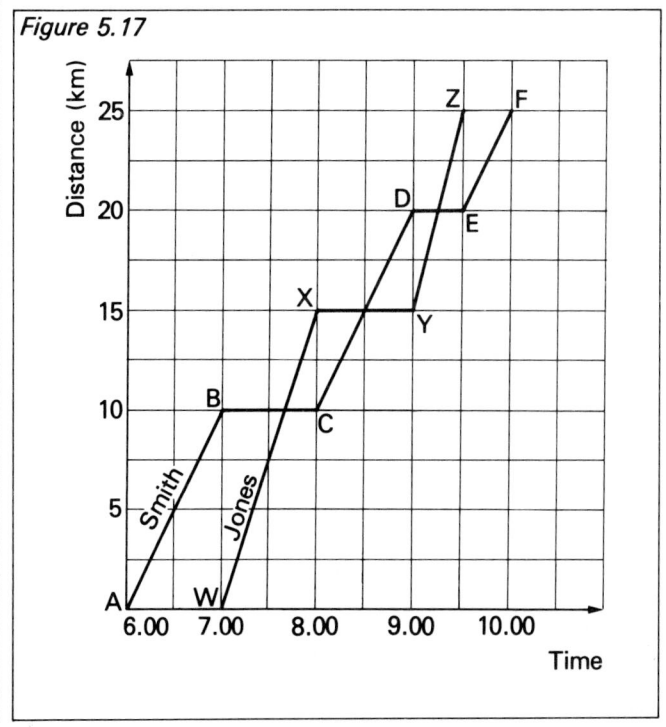

Figure 5.17

6 Linear programming

LINEAR PROGRAMMING IS A TECHNIQUE USED WIDELY IN MODERN INDUSTRY TO HELP DECISION-MAKING WHEN A VARIETY OF CONDITIONS ARE IMPOSED. IT IS POSSIBLE TO ILLUSTRATE THIS TECHNIQUE TO A LESSER DEGREE WITH SIMPLE PROBLEMS.

Example 1

A small company produces chairs and tables. The chairs cost £50 each and the tables £120 each. Every week c chairs and t tables are produced to a total value of no more than £600. Explain why

$$5c + 12t \leqslant 60$$

If at least twice as many chairs as tables are produced, complete the inequality

$$c \geqslant$$

The company undertakes to produce at least one table per week. Write an inequation to explain this statement. Now plot all three inequations on one graph and shade in the unwanted regions. Mark with a cross all the possible production alternatives. Which of these alternatives will yield the greatest amount each week?

Solution

c chairs at £50 each means a cost of £50c.
t tables at £120 each means a cost of £120t.
Since the total cost of chairs and tables cannot exceed £600

then $50c + 120t \leqslant 600$
hence $5c + 12t \leqslant 60$ (divide through by 10)

If there are at least 2 chairs built for every table

then $c \geqslant 2t$

If the company must produce at least one table per week

then $t \geqslant 1$

Figure 6.1 shows the graph of the three inequalities. Each cross represents a different solution. Point A shows the production alternative of 2 tables and 7 chairs which yields the maximum amount of £590 each week.

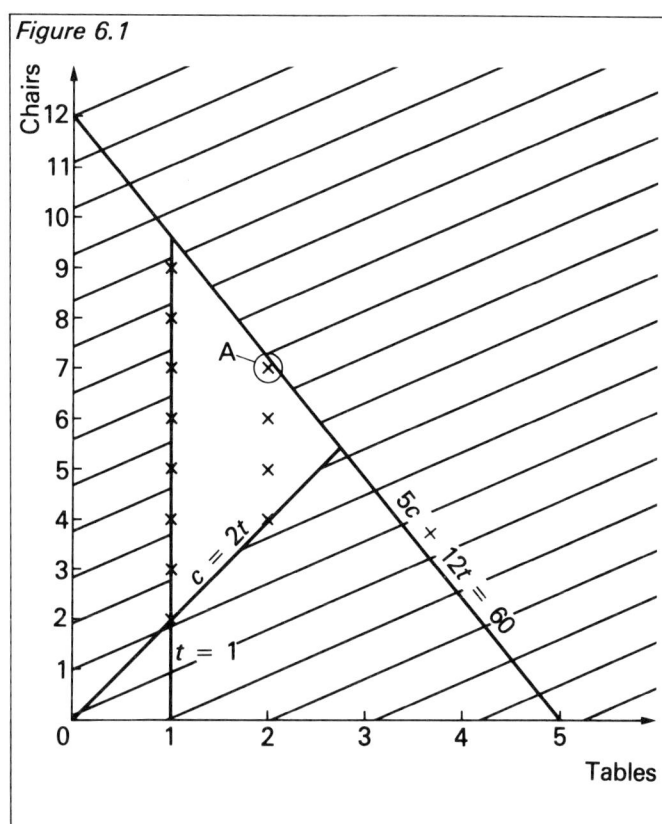

Figure 6.1

Exercise 6.1

1 Jenny has at most £30 to spend on prizes for the youth club raffle. She chooses £5 record tokens and £2 book tokens. If the number of record tokens is given by r and the number of book tokens is given by b, explain why

$$5r + 2b \leqslant 30$$

If Jenny decides to buy at least 3 record tokens, complete the inequation

$$r \geqslant$$

Plot both inequalities, shading out the regions not required. Show each ordered pair belonging to the solution set by a cross. How many different combinations are there if 4 record tokens are chosen? If 5 record tokens are chosen what is the maximum amount of money that Jenny can spend?

2 A chocolate manufacturer makes a £5 and a £3 box of chocolates, producing at most £60 worth of products every minute depending on the efficiency of the machinery. If x is the number of £5 boxes and y the number of £3 boxes produced each minute, explain why

$5x + 3y \leqslant 60$

However, because of their popularity the manufacturer always makes at least twice as many £3 boxes as £5 boxes. Complete the inequation

$y \geqslant$

Plot both inequalities, shading out the regions not required, and show all production alternatives with a cross. If the machinery is running at maximum efficiency what combination of £3 and £5 boxes of chocolates are produced?

3 On his way to school every day Andrew calls in at his local corner shop for some sweets. He likes chews at 5p each and sherbets at 8p each but he never has more than 40p to spend each day. If c is the number of chews he buys and s is the number of sherbets, complete the inequation

$5c +$ $\leqslant 40$

Andrew never buys more than 3 sherbets. Find an inequation to explain this statement.

Plot both inequations on the same graph and shade in the unwanted regions. Show all the various combinations of chews and sherbets which Andrew may choose with a cross. What is the largest number of sweets which he may buy if he has an equal number of chews and sherbets?

4 A plumber fits either 8 radiators into a bungalow or 6 radiators into a house. Let the number of bungalows he completes each week be g and the number of houses h. If he is contracted to fit at least 24 radiators each week explain why

$8g + 6h \geqslant 24$

If 48 radiators is the maximum number that he can possibly fit each week then find a second inequation to represent this information.

Now plot both inequations and shade the unwanted regions. Show all the combinations of bungalows and houses in the solution set with crosses. What is the largest number of properties that the plumber can complete in a week?

5 A clock manufacturer produces grandfather clocks at £800 each and mantel clocks at £200 each. There is a weekly production target of £6400 worth of clocks which is never exceeded. If g is the number of grandfather clocks produced each week and m the number of mantel clocks then explain why

$4g + m \leqslant 32$

Because of demand, at least 3 grandfather clocks are made each week. Complete the following

$\geqslant 3$

If it is also impossible to make more than 10 mantel clocks each week then find an inequation to explain this statement.

Plot the three inequations on one graph, shading in the unwanted regions. Mark with a cross all the production alternatives for the week. What combinations of clocks would give maximum production?

6 A pet shop keeps cats and dogs. It costs £3 a week to keep a dog and £2 a week to keep a cat. It is unprofitable for the shop to spend more than £48 a week on housing these pets. If d is the number of dogs and c the number of cats which can be kept, then explain why

$3d + 2c \leqslant 48$

Write an expression to explain the fact that the shop never has more than 12 cats.

Write a similar expression to explain the fact that the shop never has less than 4 dogs.

Plot the three inequations on one graph shading the unwanted regions and marking with a cross all the possible combinations of dogs and cats. If dogs make a profit of £7 and cats £10, use your graph to find the best combination of dogs and cats needed to make the greatest profit.

7 A school play sells tickets for adults for 50p and tickets for children for 30p. Let a be the number of 'adult' tickets sold and c the number of 'child' tickets sold. Because of a limit on seating capacity the maximum amount of money which can be made is £75. Explain why

$5a + 3c \leqslant 750$

If for every adult who attends there are 3 or less children attending then complete

$c \leqslant$

Plot these two inequations on the same graph and shade out the unwanted regions. What is the maximum number of chairs required for any night when all the available tickets are sold?

8 A van can hold a total weight of 200 kg. It normally transports two types of box, one weighing 8 kg and the other 5 kg. Let there be h, 8 kg boxes and l, 5 kg boxes. Complete the inequation

$$\leqslant 200$$

Because of distribution problems there must always be at least twice as many 8 kg boxes as 5 kg boxes. Complete the inequation

$$h \geqslant$$

There must also never be less than four 5 kg boxes. Complete the inequation

$$l \geqslant$$

Plot these three inequations on a graph, shading in all unwanted regions. Show all the combinations of the various boxes with a cross. What is the largest number of boxes that can be carried in the van? Which combination of boxes fill the van to capacity?

9 Phil Phlab likes crisps and chocbars. Every week he eats c packets of crisps and b chocbars, spending at most £4.80. If crisps cost 16p and chocbars 20p, explain why

$$4c + 5b \leqslant 120$$

Phil always buys at least 4 chocbars each week and at least twice as many crisps as chocbars. Write down two separate inequations to explain each part of the last statement.

Plot each inequation on the same graph shading the unwanted regions and show all the ordered pairs belonging to the solution set with a cross. Use your graph to find the maximum number of items which Phil Phlab can buy provided that he does not spend all his money.

10 A radio manufacturer makes two models, a superior costing £80 and a mini costing £30. Let s be the number of superior models and m the number of mini models made by a worker every hour. If the total value of the products made cannot exceed £480 an hour, explain why

$$8s + 3m \leqslant 48$$

However each worker is contracted to make no less than £240 worth of products each hour. Write down an inequation to explain this statement. Because of popular demand at least 6 mini radios must be made each hour. Complete the following inequation

$$m \geqslant$$

Plot all three inequations on the same graph, shading in the unwanted regions. Show all the production alternatives with crosses. What is the least number of radios that can be made by each worker every hour? If a mini model makes £5 profit and a superior model £15 profit, what combination of radios would the manufacturer prefer its workers to produce?

7 Geometry 2

Similar and congruent triangles

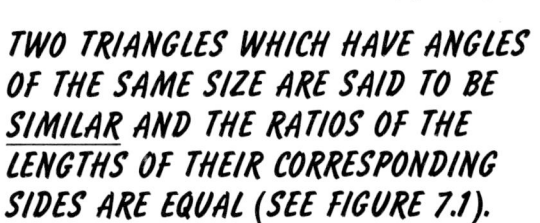

TWO TRIANGLES WHICH HAVE ANGLES OF THE SAME SIZE ARE SAID TO BE <u>SIMILAR</u> AND THE RATIOS OF THE LENGTHS OF THEIR CORRESPONDING SIDES ARE EQUAL (SEE FIGURE 7.1).

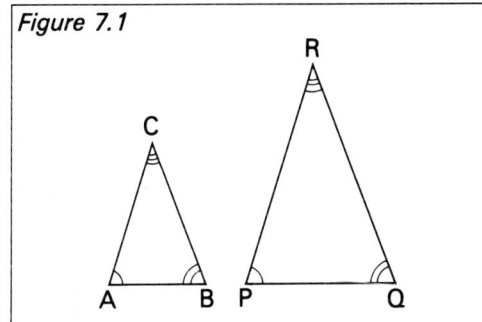

Figure 7.1

$$\angle A = \angle P, \quad \angle B = \angle Q, \quad \angle C = \angle R$$

thus $\dfrac{AC}{PR} = \dfrac{AB}{PQ} = \dfrac{BC}{QR}$

The ratio of the *areas* of two similar triangles is equal to the ratio of the square of the lengths of two corresponding sides. Thus in Figure 7.1

$$\frac{\text{area } \triangle ABC}{\text{area } \triangle PQR} = \frac{AB^2}{PQ^2} = \frac{AC^2}{PR^2} = \frac{BC^2}{QR^2}$$

TO PROVE THAT TWO TRIANGLES ARE SIMILAR, ONE OF THE FOLLOWING SETS OF CONDITIONS MUST BE SATISFIED.

(a) Two angles in one triangle must equal two angles in the second triangle. The third angle in both triangles must automatically be the same.
(b) The ratio of the lengths of all the corresponding sides of both triangles must be the same.
(c) The ratio of the lengths of two pairs of corresponding sides must be the same, plus the angle between these sides.

Example 1

In Figure 7.2 the two triangles are similar. Find the missing lengths x and y.

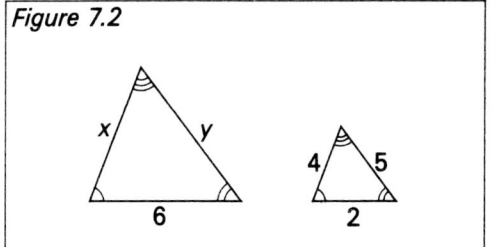

Figure 7.2

Solution

By similar triangles

$$\frac{x}{4} = \frac{y}{5} = \frac{6}{2}$$

Taking $\dfrac{x}{4} = \dfrac{6}{2}$

$\dfrac{x}{4} = 3$

hence $x = 12$

Taking $\dfrac{y}{5} = \dfrac{6}{2}$

$\dfrac{y}{5} = 3$

hence $y = 15$

Example 2

In Figure 7.3, PQ is parallel to ST. Find the lengths of y and z. If triangle STR has an area of $54\,\text{cm}^2$ find the area of triangle PQR.

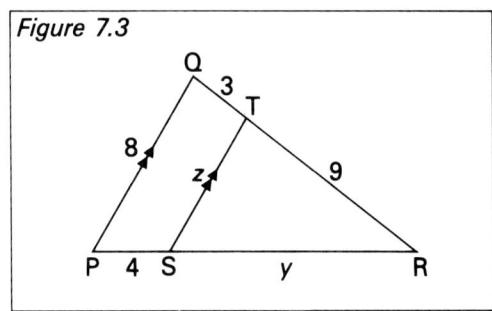

Figure 7.3

Solution

First prove that triangles STR and PQR are similar.

$\angle$R is common to both triangles.

$\angle$QPR = $\angle$TSR   (corresponding angles, since PQ parallel to ST).

Hence $\triangle$QPR is similar to $\triangle$TSR.

By similar triangles

$$\frac{QR}{TR} = \frac{PQ}{ST} = \frac{PR}{SR}$$

so $$\frac{3+9}{9} = \frac{8}{z} = \frac{4+y}{y}$$

Taking $$\frac{3+9}{9} = \frac{8}{z}$$

$$\frac{12}{9} = \frac{8}{z}$$

$$\frac{4}{3} = \frac{8}{z}$$

The numerator on the right-hand side (RHS) is twice the numerator on the left-hand side (LHS). It follows that the denominator on the RHS must also be twice the denominator on the LHS.

So $$\frac{4}{3} = \frac{8}{6}$$

(with ×2 arrows on numerator and denominator)

hence $z = 6$

Taking $$\frac{3+9}{9} = \frac{4+y}{y}$$

$$\frac{12}{9} = \frac{4+y}{y}$$

$$\frac{4}{3} = \frac{4+y}{y}$$

Find the common denominator of both fractions, thus

$$3 \times y = 3y$$

Now multiply both sides by the common denominator

$$3y \times \frac{4}{3} = 3y \times \left(\frac{4+y}{y}\right)$$

Simplify both sides

$$y \times 4 = 3 \times (4+y)$$
$$4y = 12 + 3y \quad \text{(subtract } 3y \text{ from both sides)}$$
$$y = 12$$

$$\frac{\text{area } \triangle PQR}{\text{area } \triangle STR} = \frac{QR^2}{TR^2}$$

so $$\frac{\text{area } \triangle PQR}{54} = \frac{12^2}{9^2}$$

$$= \frac{144}{81}$$

$$= \frac{16}{9} \quad \text{(divide top and bottom by 9)}$$

$$54 \times \frac{\text{area } \triangle PQR}{54} = 54 \times \frac{16}{9} \quad \text{(multiply both sides by 54 and simplify)}$$

$$\text{area } \triangle PQR = 6 \times 16$$
$$= 96 \, cm^2$$

TWO TRIANGLES ARE <u>CONGRUENT</u> ONLY WHEN THEY ARE THE <u>SAME IN EVERY RESPECT</u>. THIS MEANS THAT ALL THE CORRESPONDING SIDES AND ANGLES OF BOTH TRIANGLES MUST BE EQUAL (SEE FIGURE 7.4).

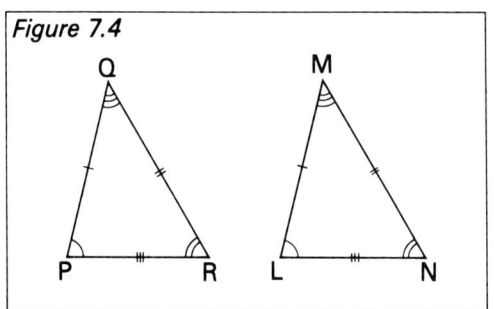

Figure 7.4

$$\angle P = \angle L, \quad \angle Q = \angle M, \quad \angle R = \angle N$$
and $\quad PQ = LM, \quad PR = LN, \quad QR = MN$

> **TO PROVE THAT TWO TRIANGLES ARE CONGRUENT, ONE OF THE FOLLOWING SETS OF CONDITIONS MUST BE SATISFIED.**

(a) The lengths of all the sides of one triangle must be equal to those of the other triangle. (SSS)
(b) The lengths of two sides and the included angle in one triangle must be equal to the lengths of two sides and the included angle in the other triangle. (SAS)
(c) Two angles and the length of one side in the first triangle must be equal to the two angles and one side in a corresponding position on the other triangle. (ASA)

Example 3

In Figure 7.5, AD is parallel to EG and BF is parallel to DG. AC = EG. $\angle BAC = \angle GEH$ and $\angle DGH = 2\angle ACB$. Find the length of GH.

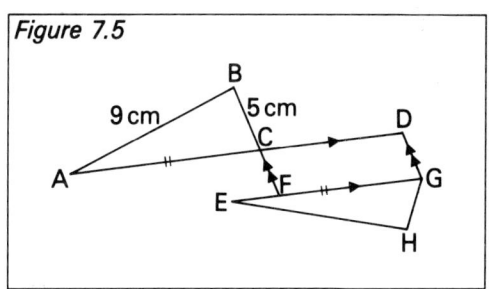

Figure 7.5

Solution

Prove that $\triangle ABC$ and $\triangle EGH$ are congruent.

$\quad \angle ACB = \angle DCF \quad$ (vertically opposite)
$\quad \angle DCF = \angle DGF \quad$ (opposite angles of
$\qquad\qquad\qquad\qquad$ parallelogram equal)
so $\quad \angle ACB = \angle DGF \qquad\qquad\qquad$ (1)

But if $\quad \angle DGF + \angle FGH = \angle DGH = 2\angle ACB$
then $\qquad \angle ACB + \angle FGH = 2\angle ACB \quad$ (from equation (1))
and $\qquad\qquad\qquad \angle FGH = \angle ACB$

$$\angle BAC = \angle GEH \quad \text{(given)}$$
$$AC = EG \quad \text{(given)}$$

so $\triangle ABC$ is congruent to $\triangle EGH$. (ASA)
This is often written $\triangle ABC \equiv \triangle EGH$.

Hence $\quad GH = 5\,cm$

Exercise 7.1

1 Find the lengths of x and y in Figure 7.6.

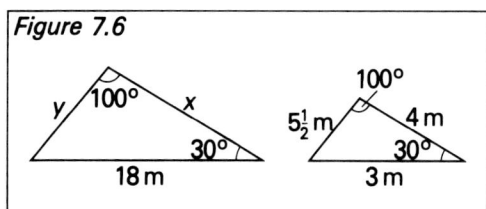

Figure 7.6

2 In Figure 7.7 two sets of angles are shown to be equal. Find the lengths of a and b.

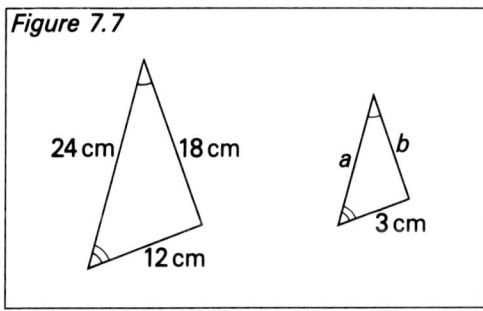

Figure 7.7

3 Figure 7.8 shows two similar triangles. Find the lengths of h and k.

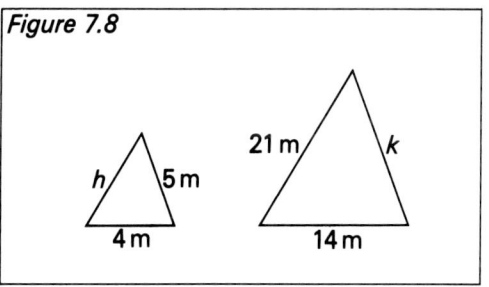

Figure 7.8

4 If the area of triangle PQR in Figure 7.9 is 10 cm², find the area of triangle XYZ.

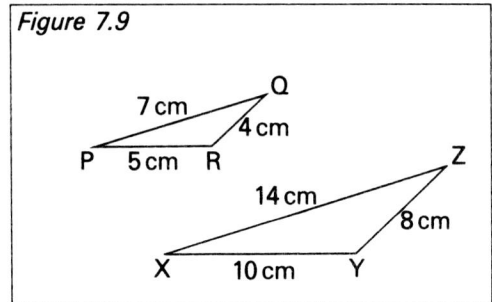

Figure 7.9

5 Find the area of triangle DEF in Figure 7.10 when the area of triangle ABC is 108 m².

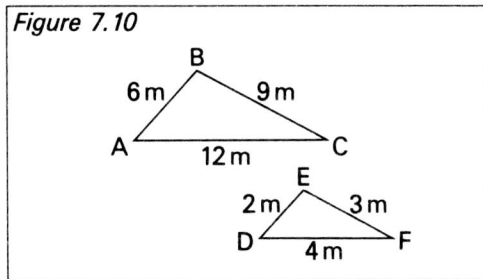

Figure 7.10

6 Which pairs of triangles in Figure 7.11 are congruent?

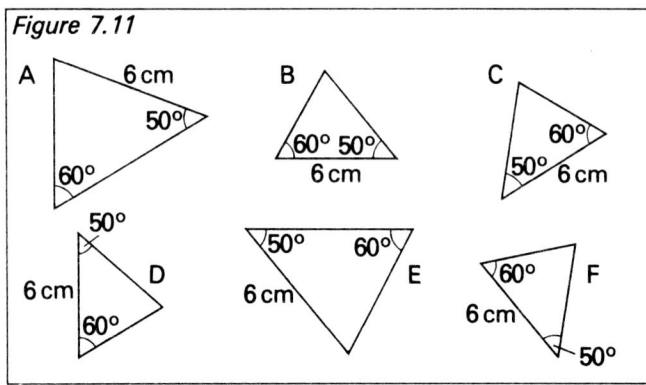

Figure 7.11

7 Find the length of BC in Figure 7.12, proving any assumptions which you make.

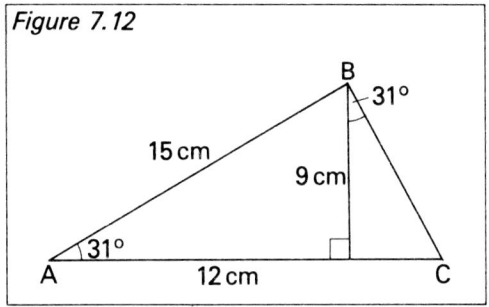

Figure 7.12

8 In Figure 7.13, prove that triangle PQT is similar to triangle RST, hence find the lengths of x and y. If the area of triangle RST is 12.5 cm², find the area of triangle PQT.

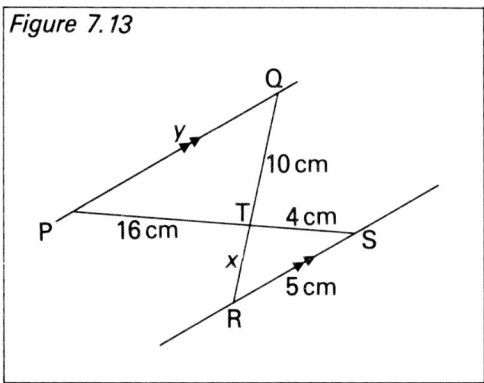

Figure 7.13

9 In Figure 7.14, PS = QS = QR and PQ = 20 cm. Find the length of SR proving any assumptions which you make.

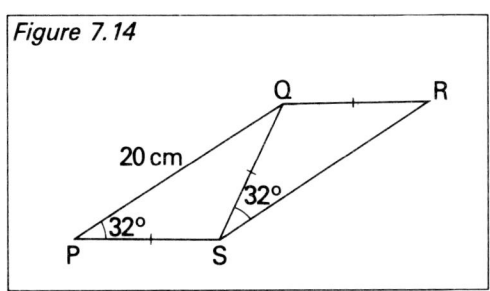

Figure 7.14

10 Prove that triangle XYZ is similar to triangle XAB in Figure 7.15. YZ = 20 cm, AB = 8 cm, XZ = 30 cm and XA = 6 cm. Find the lengths of XY and XB. Given that the area of triangle XYZ is 125 cm², find the area of triangle XAB.

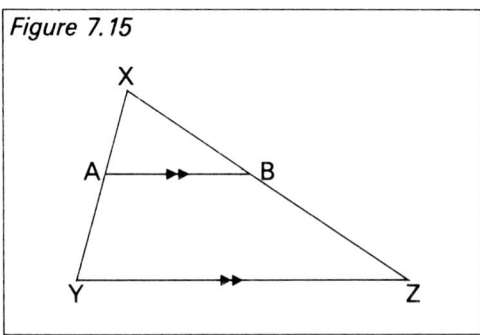

Figure 7.15

11 In Figure 7.16, ∠BAF = 41° and ∠DFE = 49°. BCDF is a square of side 5 cm and AB = 6 cm. Find the length of DE proving any assumptions which you make.

12 Find the lengths of x and y in Figure 7.17 proving any assumptions which you make. Find the area of triangle ACD given that the area of triangle ABD is 20 cm².

60

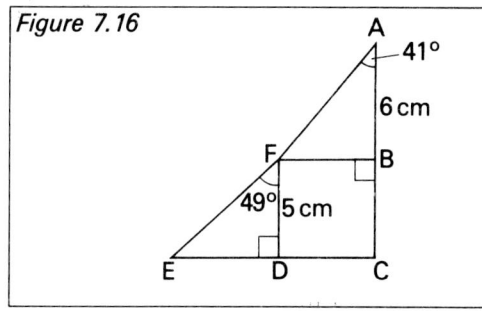

Figure 7.16

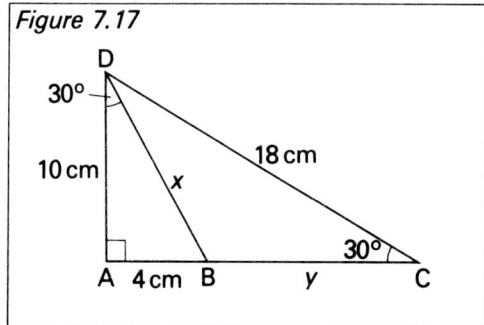

Figure 7.17

Angle properties of the circle

Definitions (see Figure 7.18)

(a) A chord is a straight line which internally connects two points on the circumference of a circle.

(b) A diameter is a special chord which passes through the centre of a circle.

(c) A radius is a straight line joining the centre of a circle to a point on the circumference.

(d) A secant is a straight line which passes through the circumference of a circle at two points.

(e) A tangent is a straight line which touches the circumference of a circle at one point only.

(f) A segment is an area within a circle enclosed by a chord and the circumference.

(g) An arc is part of the circumference of a circle.

(h) A sector is part of a circle bounded by two radii and the subtended arc.

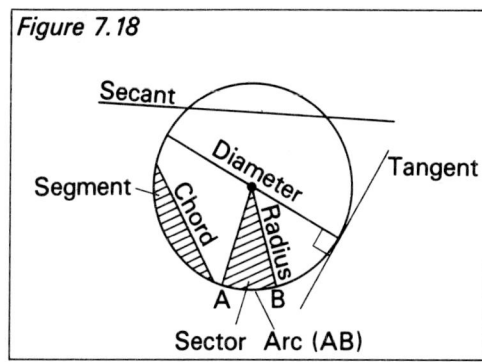

Figure 7.18

Theorem 1 (Figure 7.19)

If a side of any cyclic triangle is a diameter then the angle opposite this diameter is always a *right angle*.

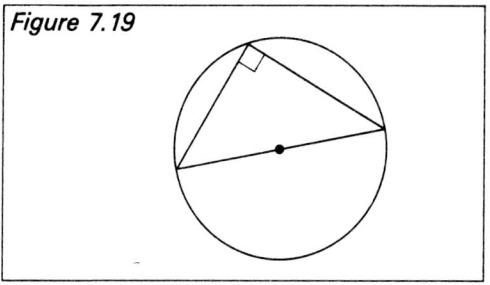

Figure 7.19

Theorem 2 (Figure 7.20)

All angles, *a*, subtended by the same arc of a circle are equal.

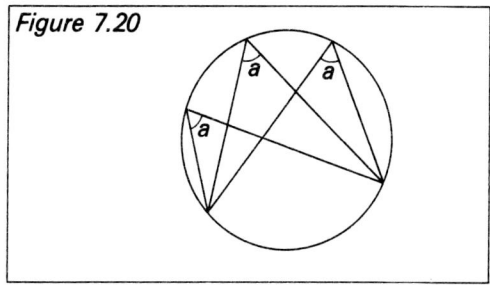

Figure 7.20

Theorem 3 (Figure 7.21)

Any angle which an arc subtends at the centre of a circle is always twice the size of the angle subtended by the same arc on the remaining part of the circumference.

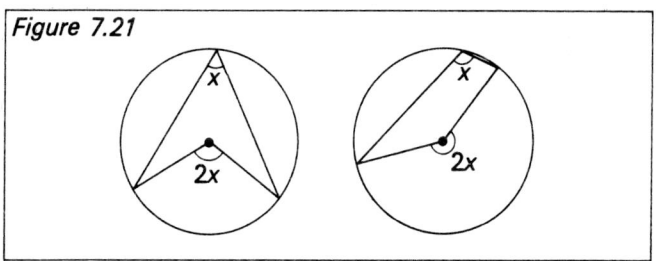

Figure 7.21

Theorem 4 (Figure 7.22)

A quadrilateral inscribed in a circle is known as a *cyclic* quadrilateral. Opposite pairs of internal angles are supplementary hence, $a+c = b+d = 180°$.

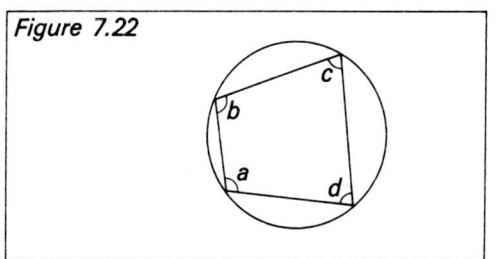

Figure 7.22

Theorem 5 (Figure 7.23)

Any radius which is drawn to the tangent of a circle is always perpendicular to it.

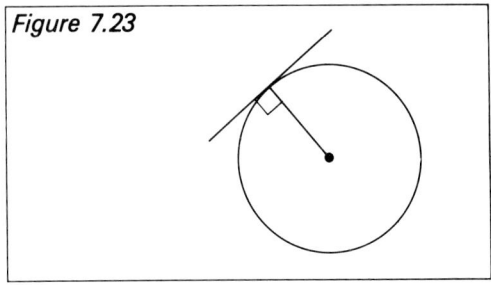
Figure 7.23

Theorem 6 (Figure 7.24)

A straight line from the centre of a circle to a point outside the circle bisects the angle made by the two tangents from the point to the circle. Also the lengths of the two tangents are equal.

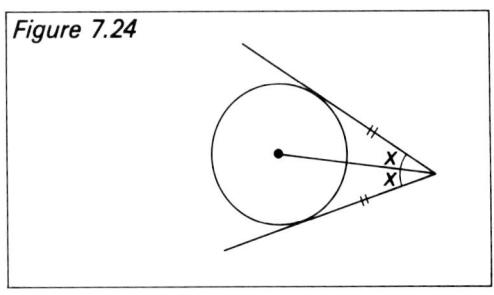
Figure 7.24

Theorem 7 (Figure 7.25)

The angle which any tangent makes with a chord is equal to the angle subtended in the alternate segment.

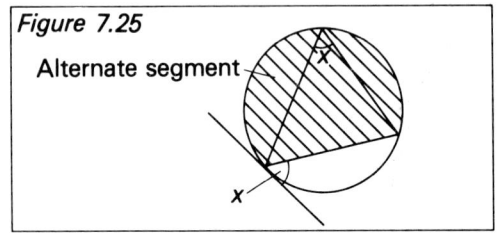
Figure 7.25
Alternate segment

Example 4

Find the missing angles indicated in Figure 7.26, giving reasons for each of the assumptions made.

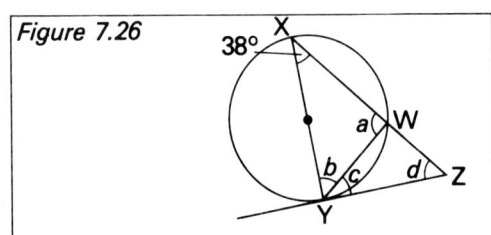
Figure 7.26

Solution

In $\triangle$WXY, $a = 90°$ (theorem 1 – angle in a semi-circle)

In $\triangle$WXY
$$a + b + 38° = 180° \text{(angles in a triangle)}$$
so $90° + b + 38° = 180°$
$$b + 128° = 180°$$
$$b = 52°$$

In $\triangle$XYZ
$$b + c = 90° \text{(theorem 5 – radius drawn to a tangent is perpendicular)}$$
so $52° + c = 90°$
$$c = 38°$$

Theorem 7 could also have been used to find c in $\triangle$XYZ.

In $\triangle$WYZ
$$38° + 90° + d = 180° \text{(angles in a triangle)}$$
$$d + 128° = 180°$$
$$d = 52°$$

Example 5

Find the missing angles in Figure 7.27, explaining your logic as the solution progresses.

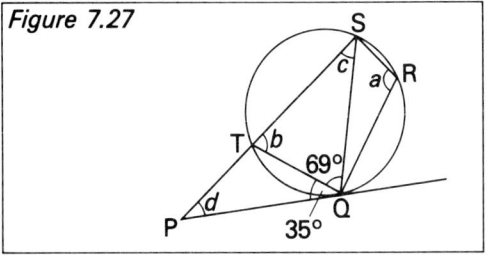
Figure 7.27

Solution

$\angle$PQS = $\angle$QRS (theorem 7 – alternate segment)
so $35° + 69° = a$
$$a = 104°$$

$\angle$QTS + $\angle$QRS = 180° (theorem 4 – opposite angles of cyclic quadrilateral are supplementary)

so $a + b = 180°$
$$104° + b = 180°$$
$$b = 76°$$

In $\triangle$QST
$$b + c + 69° = 180° \text{(angles in a triangle)}$$
$$76° + c + 69° = 180°$$
$$c + 145° = 180°$$
$$c = 35°$$

In △PQS

$c + d + 35° + 69° = 180°$ (angles in a triangle)
$35° + d + 35° + 69° = 180°$
$d + 139° = 180°$
$d = 41°$

Exercise 7.2

In each question, state clearly any assumptions which you make.

1 Find angles a and b in Figure 7.28.

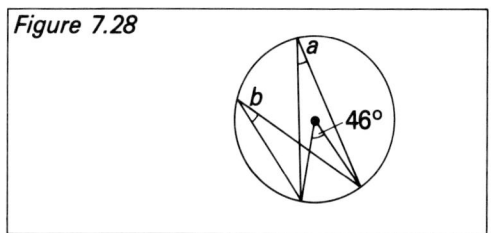

Figure 7.28

2 Find the lettered angles in Figure 7.29.

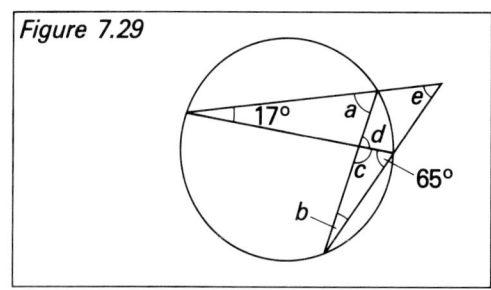

Figure 7.29

3 Find the missing angles a, b and c in Figure 7.30.

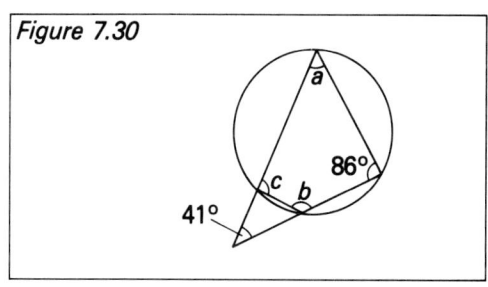

Figure 7.30

4 In Figure 7.31, ∠BAD = 54°, BC is a diameter and DF a radius. Find   (a) ∠BAF   (b) ∠AFD   (c) ∠DFE   (d) ∠AEF.

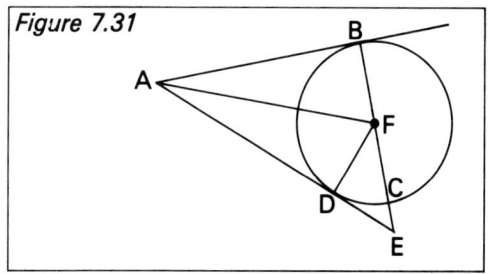

Figure 7.31

5 In Figure 7.32, BE is a diameter, ∠CBE = 63°, ∠AEB = 36°, ∠DCE = 37°. Find the following angles.   (a) ∠BCE   (b) ∠BEC   (c) ∠CED   (d) ∠CDE   (e) ∠BAE   (f) ∠ABC   (g) ∠DEF   (h) ∠DFE.

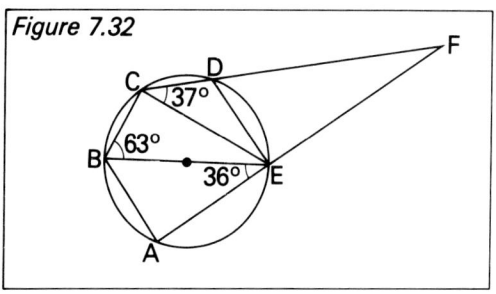

Figure 7.32

6 In Figure 7.33, AC is a diameter, ∠BAD = 110° and ∠CBD = 73°. Find the missing angles indicated.

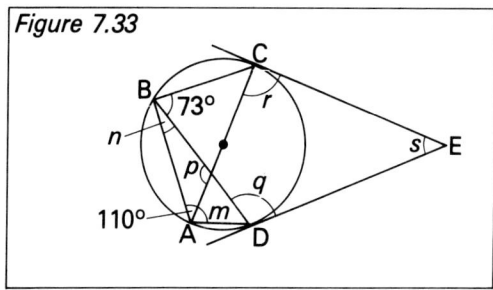

Figure 7.33

7 Find the missing lettered angles in Figure 7.34.

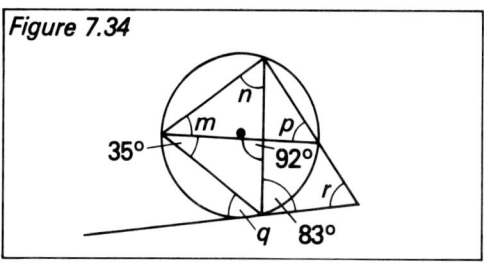

Figure 7.34

8 In Figure 7.35, ∠ABG = 40°, ∠AGD = 95° and ∠DFE = 38°. Find the following angles.
(a) ∠ACD   (b) ∠BAG   (c) ∠EAG   (d) ∠CDG
(e) ∠AED   (f) ∠CDE   (g) ∠EDG.

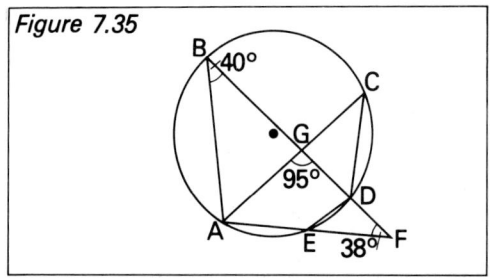

Figure 7.35

9 Find the missing angles in Figure 7.36 given that AB is a diameter of the circle, ∠CBE = 22° and ∠BED = 38°.

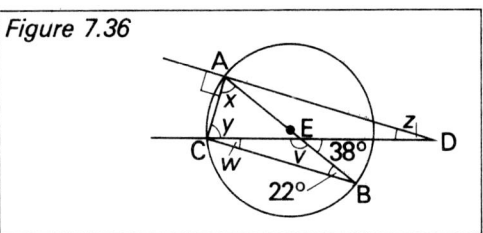

Figure 7.36

10 In Figure 7.37, AC and BD are diameters of the circle and the tangent T touches the circle at A such that ∠TAB = 70°. Find the following angles.
(a) ∠ADB   (b) ∠BAC   (c) ∠ABD   (d) ∠ACD.

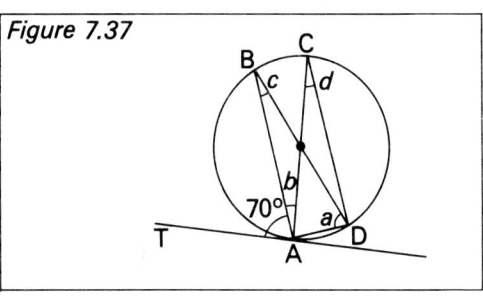

Figure 7.37

64

8 Flow charts

> A *FLOW CHART* GIVES A STEP-BY-STEP PROCEDURE FOR SOLVING A GIVEN MATHEMATICAL PROBLEM. THIS TYPE OF PROCEDURE IS OFTEN CALLED AN *ALGORITHM*.

Example 1

Follow the flow chart in Figure 8.1 and complete the trace table when the input numbers are
(a) $P = 8$, $Q = 5$ (b) $P = 13$, $Q = -3$
(c) $P = -2$, $Q = 6$.

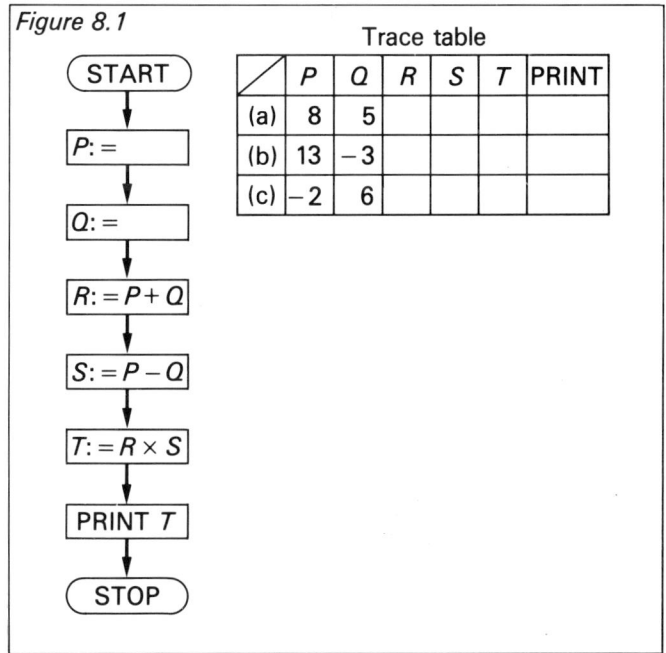

Figure 8.1

Trace table

	P	Q	R	S	T	PRINT
(a)	8	5				
(b)	13	−3				
(c)	−2	6				

Solution

Write the intermediate values for R, S and T in the trace table from the instructions given in the flow chart. Finally finish the trace table by writing in the value for PRINT T. (See Figure 8.2.)

Figure 8.2

	P	Q	R	S	T	PRINT
(a)	8	5	13	3	39	39
(b)	13	−3	10	16	160	160
(c)	−2	6	4	−8	−32	−32

Example 2

Follow the flow chart in Figure 8.3 and complete the trace table when the input numbers are $A = 3$ and $B = 12$.

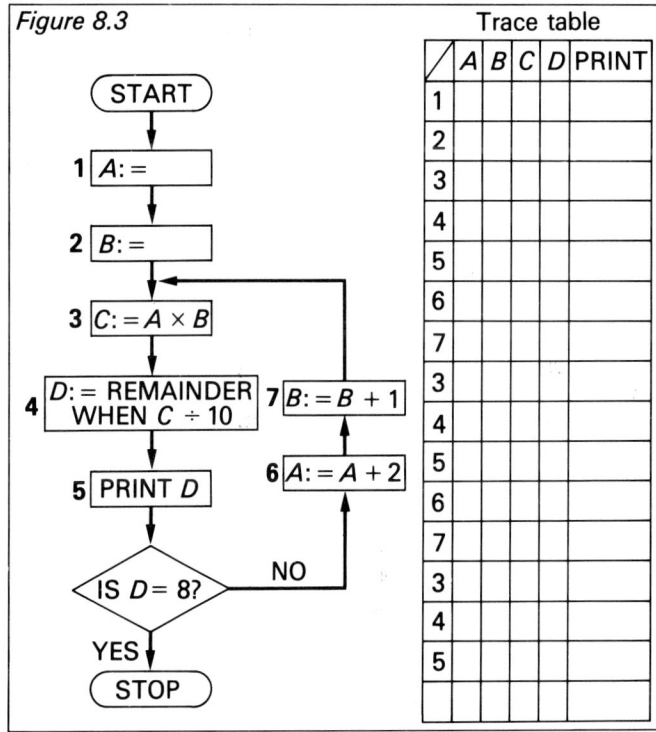

Figure 8.3

Trace table

	A	B	C	D	PRINT
1					
2					
3					
4					
5					
6					
7					
3					
4					
5					
6					
7					
3					
4					
5					

Solution

Follow the flow chart through. Each time you reach a numbered box, carry out the instruction inside it and record the result on to the trace table.

IF THE ANSWER TO THE DECISION BOX IS 'NO', CARRY ON ROUND THE LOOP AND COMPLETE THE TRACE TABLE FURTHER. KEEP GOING AROUND THE LOOP UNTIL THE ANSWER IN THE DECISION BOX IS 'YES'. THIS WILL LEAD YOU TO 'STOP' AND THE PROBLEM IS THEN COMPLETED.

(See Figure 8.4 for the completed trace table.)

Figure 8.4

	A	B	C	D	PRINT
1	3				
2	3	12			
3	3	12	36		
4	3	12	36	6	
5	3	12	36	6	6
6	5	12	36	6	
7	5	13	36	6	
3	5	13	65	6	
4	5	13	65	5	
5	5	13	65	5	5
6	7	13	65	5	
7	7	14	65	5	
3	7	14	98	5	
4	7	14	98	8	
5	7	14	98	8	8
	S	T	O	P	

Exercise 8.1

In questions 1 to 4 complete the trace tables for the flow charts and input information given.

1

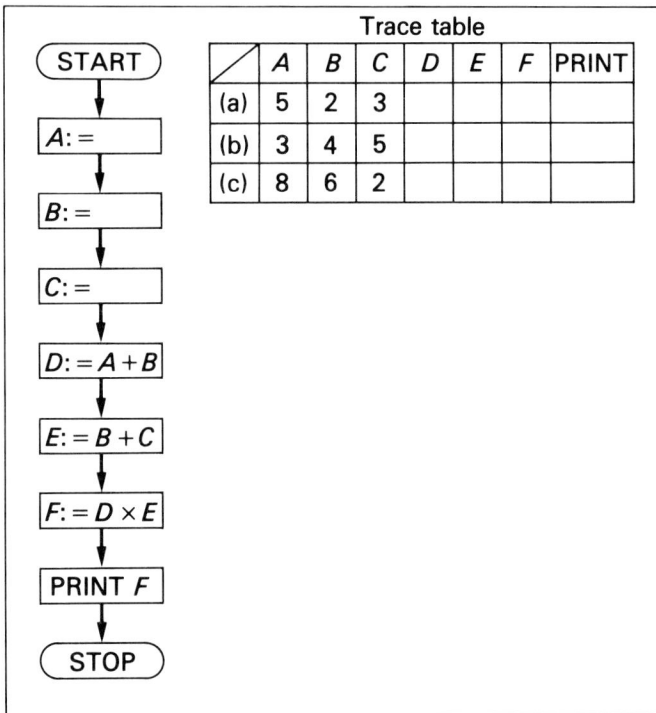

Trace table

	A	B	C	D	E	F	PRINT
(a)	5	2	3				
(b)	3	4	5				
(c)	8	6	2				

2

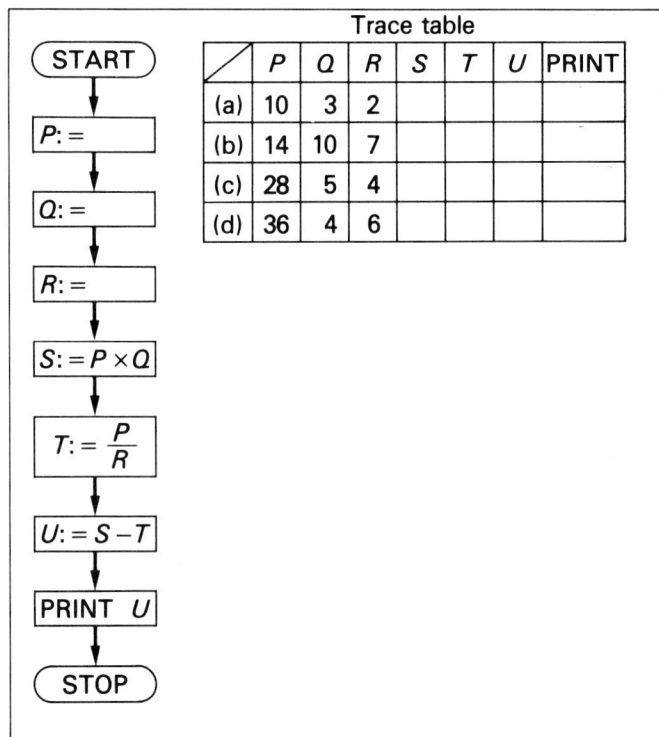

Trace table

	P	Q	R	S	T	U	PRINT
(a)	10	3	2				
(b)	14	10	7				
(c)	28	5	4				
(d)	36	4	6				

3

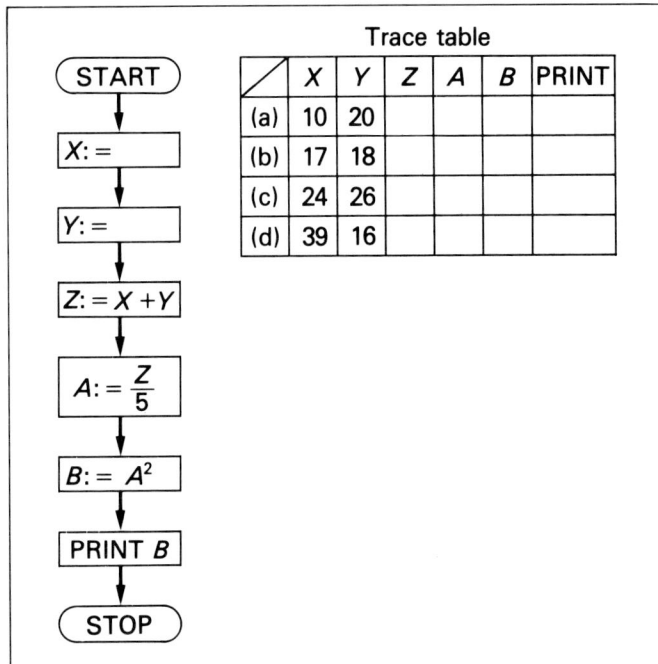

Trace table

	X	Y	Z	A	B	PRINT
(a)	10	20				
(b)	17	18				
(c)	24	26				
(d)	39	16				

4

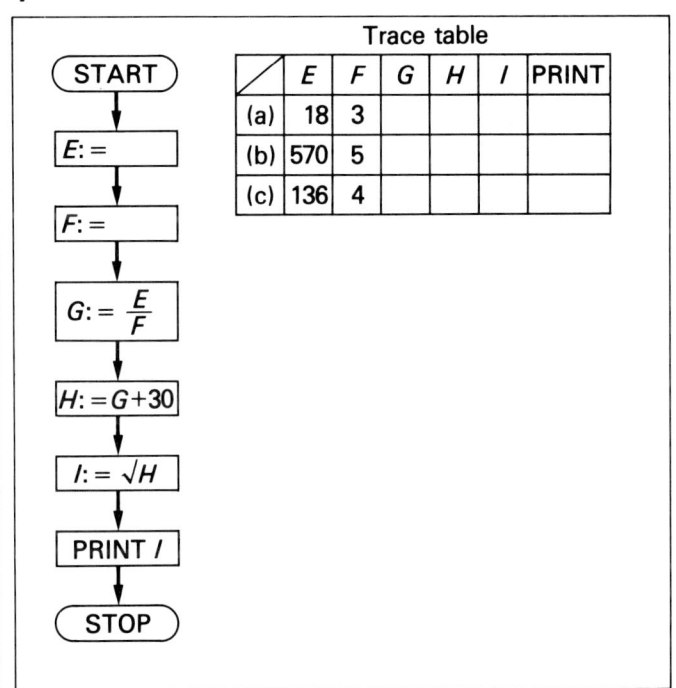

Trace table

	E	F	G	H	I	PRINT
(a)	18	3				
(b)	570	5				
(c)	136	4				

5 Work through the flow chart in Figure 8.5 and complete the trace table. Name the series of values in the print column.

6 Work through the flow chart in Figure 8.6 and complete a trace table for the following.
(a) $A = 3$, $B = 2$ (b) $A = 2$, $B = 4$
(c) $A = 7$, $B = -2$.

Figure 8.5

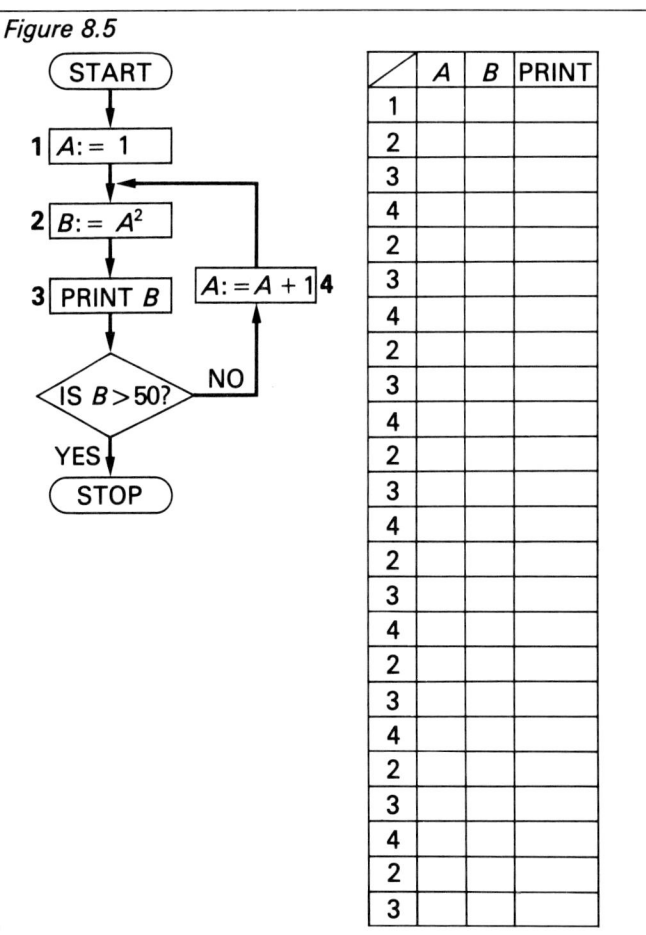

	A	B	PRINT
1			
2			
3			
4			
2			
3			
4			
2			
3			
4			
2			
3			
4			
2			
3			
4			
2			
3			
4			
2			
3			

Figure 8.6

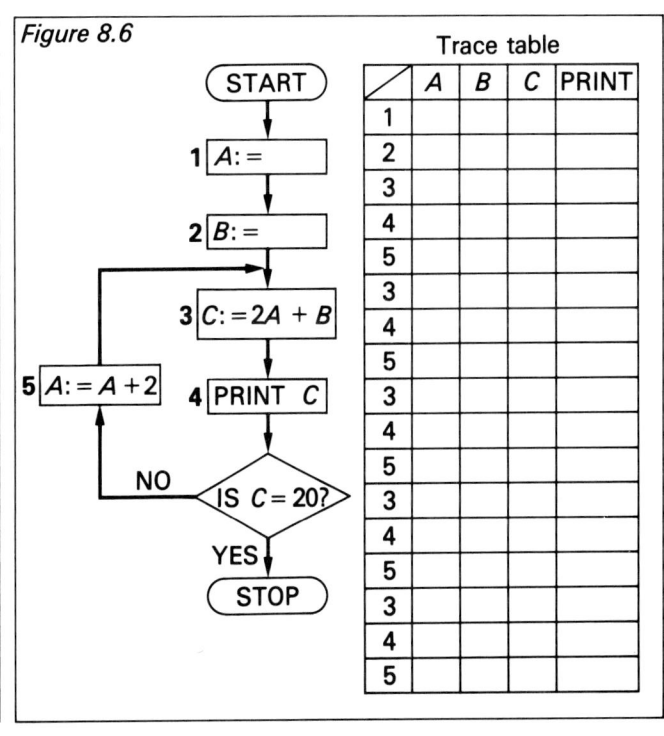

Trace table

	A	B	C	PRINT
1				
2				
3				
4				
5				
3				
4				
5				
3				
4				
5				
3				
4				
5				
3				
4				
5				

7 Work through the flow chart in Figure 8.7 and complete a trace table for the following.
(a) $A = 25$, $B = 3$ (b) $A = -2$, $B = 10$
(c) $A = -3$, $B = -2$.

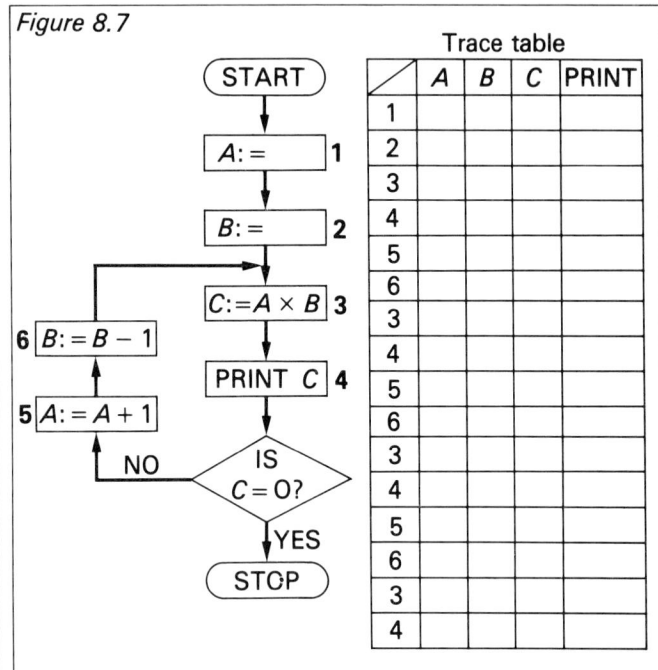

Figure 8.7

	A	B	C	PRINT
1				
2				
3				
4				
5				
6				
3				
4				
5				
6				
3				
4				
5				
6				
3				
4				

Trace table

8 Work through the flow chart in Figure 8.8 and complete a trace table for the following.
(a) $X = 2$, $Y = 3$ (b) $X = 4$, $Y = 2$.

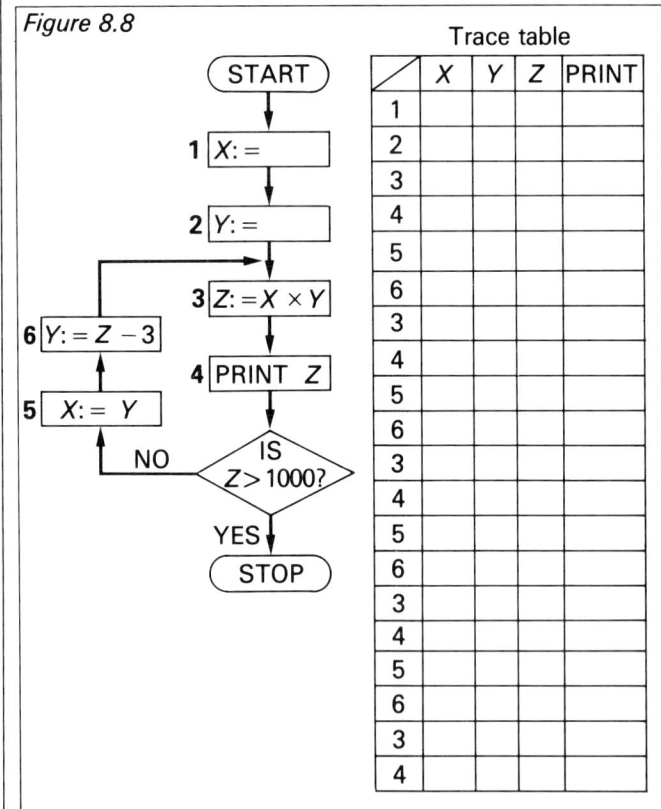

Figure 8.8

	X	Y	Z	PRINT
1				
2				
3				
4				
5				
6				
3				
4				
5				
6				
3				
4				
5				
6				
3				
4				
5				
6				
3				
4				

Trace table

9 Work through the flow chart in Figure 8.9 and complete a trace table for the following.
(a) $A = 10$ (b) $A = 15$ (c) $A = 22$.
   Try this flow chart for a positive whole number of your choice. Will the flow chart work for any positive whole number?

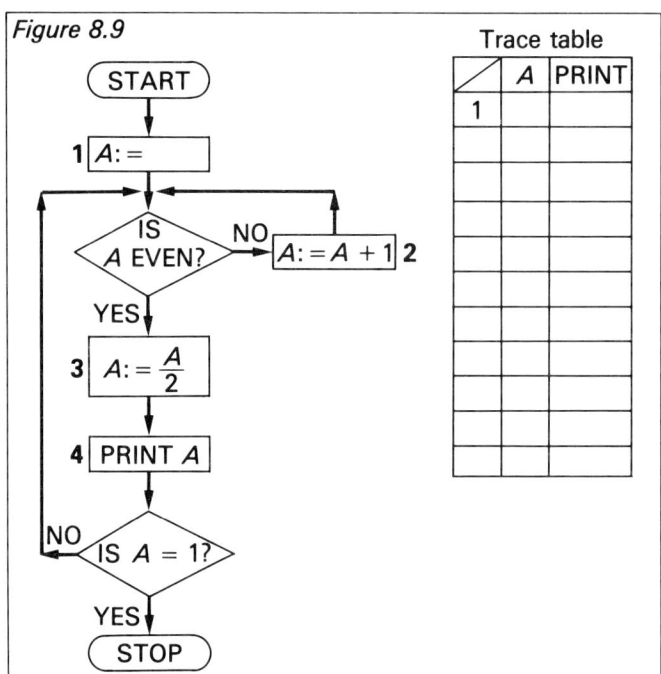

Figure 8.9

	A	PRINT
1		

Trace table

10 Complete the trace table for the flow chart in Figure 8.10. Can you name the series found in the print column of the trace table?

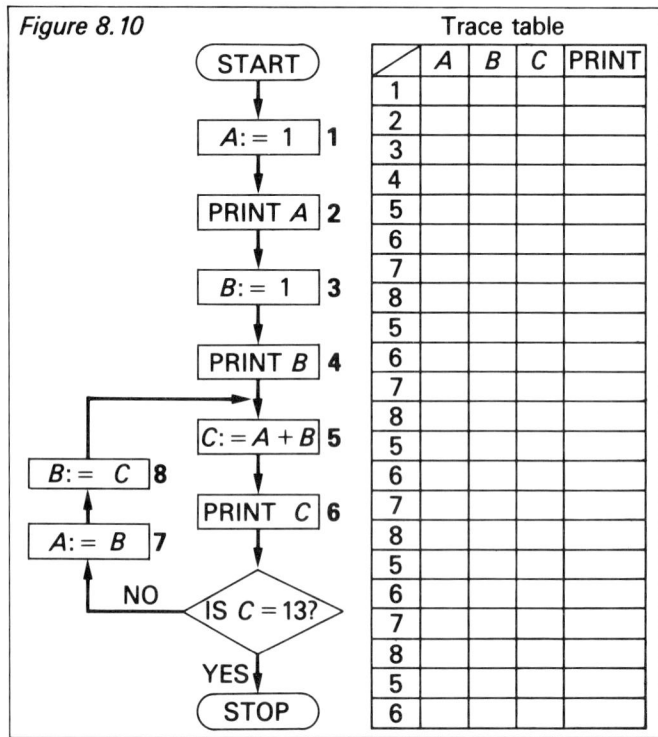

Figure 8.10

	A	B	C	PRINT
1				
2				
3				
4				
5				
6				
7				
8				
5				
6				
7				
8				
5				
6				
7				
8				
5				
6				

Trace table

9 Trigonometry 2

Solid shapes

These questions involve finding lengths and angles of right-angled triangles within three-dimensional shapes. The Theorem of Pythagoras is also widely used (see Book 1, chapter 18).

> *IT IS VITALLY IMPORTANT TO DRAW GOOD DIAGRAMS WHEN ANSWERING THIS TYPE OF QUESTION SINCE THEY OFTEN HELP YOU MAKE DECISIONS ABOUT HOW TO TACKLE A PARTICULAR PROBLEM.*

Example 1

Figure 9.1 shows a door wedge.
Calculate   (a) the length of the sloping edge of the wedge   (b) the diagonal AF of the base rectangle
(c) $\angle$ EAF.

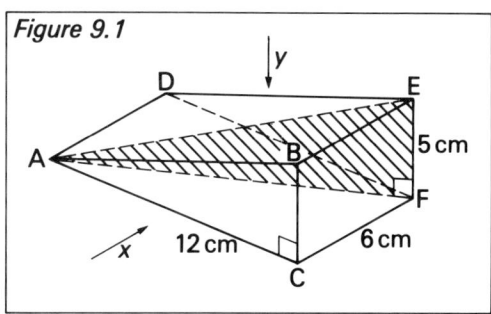

Figure 9.1

Solution

(a) View the wedge in direction x (see Figure 9.2).

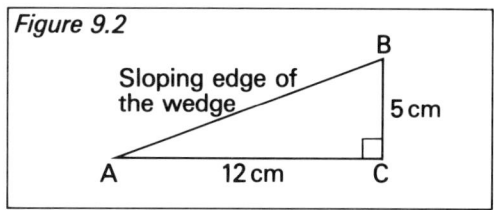

Figure 9.2

Using the Theorem of Pythagoras

$$\begin{aligned}
AB^2 &= BC^2 + AC^2 \\
&= 5^2 + 12^2 \\
&= 25 + 144 \\
&= 169 \\
AB &= \sqrt{169} = 13\,cm
\end{aligned}$$

(b) View the wedge in direction y (see Figure 9.3).

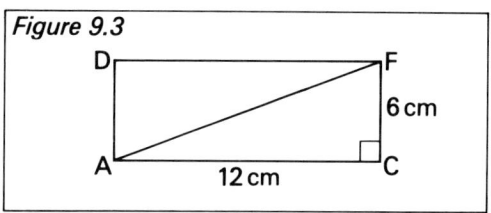

Figure 9.3

Using the Theorem of Pythagoras

$$\begin{aligned}
AF^2 &= AC^2 + FC^2 \\
&= 12^2 + 6^2 \\
&= 144 + 36 \\
&= 180 \\
AF &= \sqrt{180} = 13.42\,cm
\end{aligned}$$

(c) Take out the shaded triangle (see Figure 9.4). Let $\angle$ EAF $= a$.

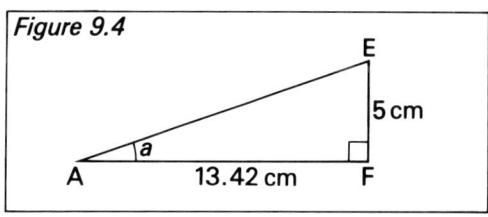

Figure 9.4

The triangle has   (i) an angle (a)
(ii) opposite side (5 cm)   (iii) adjacent side (12.65 cm), so use tangent ratio

$$\tan a = \frac{\text{opposite side}}{\text{adjacent side}}$$

$$\tan a = \frac{5}{13.42}$$

$$\begin{aligned}
&= 0.3726 \quad \text{(using logs or a calculator)} \\
a &= 20° 26'
\end{aligned}$$

Exercise 9.1

1 Calculate the lengths of AC and AX in Figure 9.5 correct to 2 decimal places. Find also angles
(a) ∠BAC (b) ∠XAB (c) ∠XCB,
giving your answers to the nearest degree.

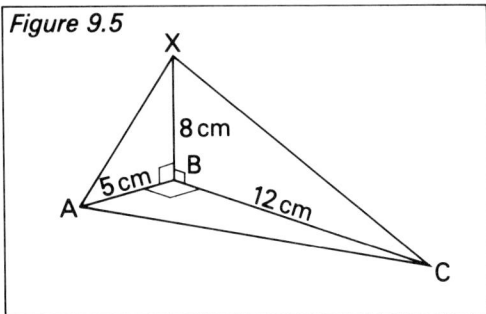

Figure 9.5

2 Figure 9.6 shows a cube of side 6 cm. Calculate
(a) length AH (b) length AG (c) ∠HAG
(d) ∠AHE. (Give both lengths correct to 2 decimal places and both angles correct to the nearest degree.)

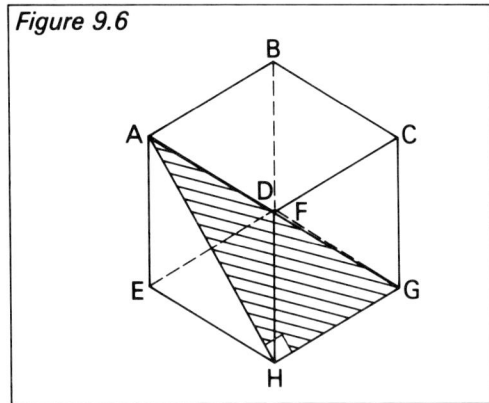

Figure 9.6

3 Figure 9.7 shows a triangular prism. Triangle ABC is isosceles having AB = AC = 13 cm and BC = 10 cm. The length of the prism is 25 cm.
Calculate (a) the vertical height (AO) of the prism (b) ∠ACO (c) ∠BAC (d) length OZ
(e) ∠AZO. (Give all angles correct to the nearest degree.)

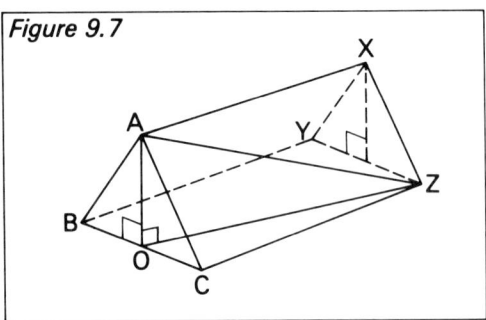

Figure 9.7

4 The pyramid in Figure 9.8 has a square base of side 12 cm and a vertical height of 8 cm. W is the midpoint of CD.
Calculate (a) length XW (b) ∠XWO
(c) length AC (d) ∠XCO. (Give both angles correct to the nearest degree.)

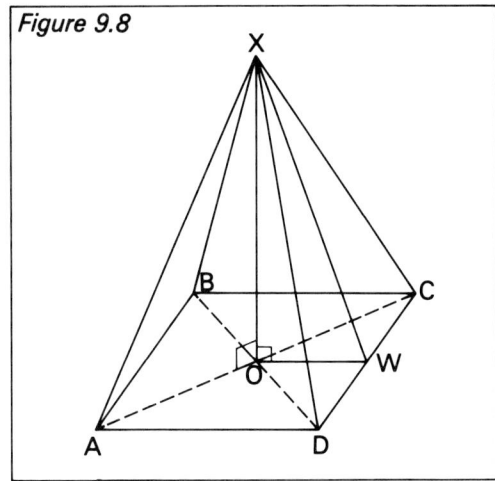

Figure 9.8

5 Figure 9.9 shows a right circular cone with a diameter of 20 cm and vertical height of 24 cm.
Calculate (a) slant height (AB) (b) apex angle (∠BAC) to the nearest degree.

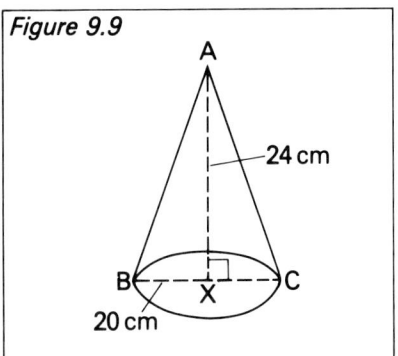

Figure 9.9

6 Figure 9.10 shows a cuboid with a square cross-section of side 9 cm and a length of 40 cm. M and N are the midpoints of BF and AE respectively.
Find (a) length EG (b) ∠GEC
(c) length CM (d) ∠CNM. (Give both lengths correct to 2 decimal places and both angles correct to the nearest degree.)

7 In Figure 9.11, find (a) length AP (b) ∠AXP
(c) angle which face AWX makes with the horizontal (d) length WY (e) ∠AWO
(f) length AW. (Give all lengths correct to 2 decimal places and all angles correct to the nearest degree.)

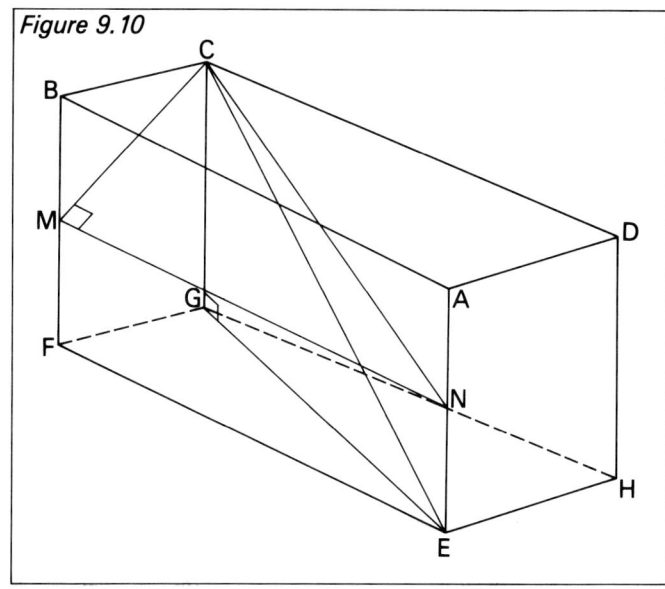

Figure 9.10

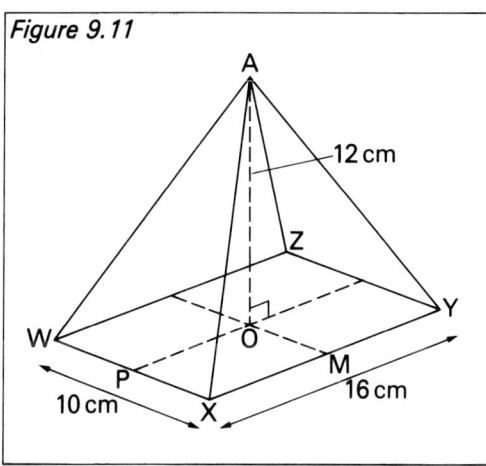

Figure 9.11

8 Using Figure 9.12, find   (a) ∠BFC
(b) length EC   (c) ∠BEC   (d) length AE
(e) ∠AEB. (Give all lengths correct to 2 decimal
places and all angles correct to the nearest degree.)

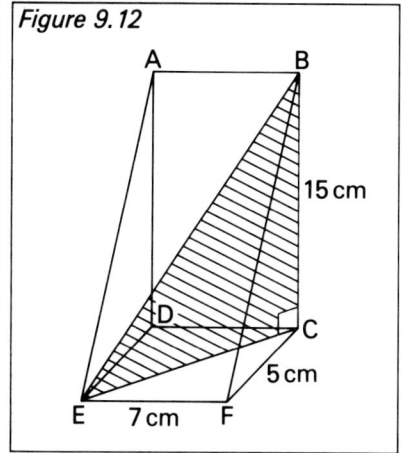

Figure 9.12

Bearing problems

Example 2

A tractor is driven on a bearing of 120° for 100 m before
travelling a further 70 m due north. The tractor is then
driven 50 m due west before coming to rest
(see Figure 9.13).

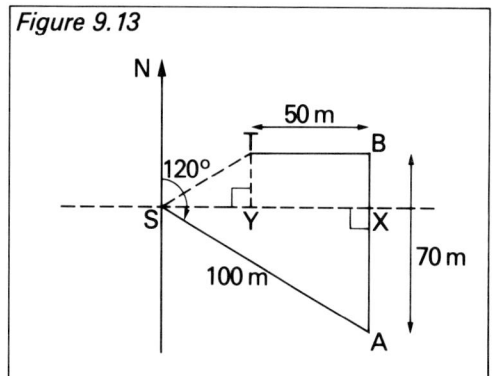

Figure 9.13

Find   (a) how far the tractor goes due south
(b) how far the tractor goes due north   (c) how far the
tractor had been driven due east when it stopped
(d) the final bearing of the tractor from its starting
position, to the nearest degree.

Solution

(a) Due south distance is XA.
Using trigonometry in △SXA

$$\sin X\widehat{S}A = \frac{XA}{SA}$$

but $X\widehat{S}A = 120° - 90° = 30°$
so $\sin 30° = \dfrac{XA}{100}$

$$\Rightarrow \quad XA = 0.5 \times 100$$
$$= 50\,m$$

(b) Due north distance is XB.

But   XB = BA − XA
= 70 m − 50 m
= 20 m

(c) Due east distance is SY.

But   SY = SX − YX

Using trigonometry in △SXA

$$\cos X\widehat{S}A = \frac{SX}{SA}$$

so $\cos 30° = \dfrac{SX}{100}$

$$\Rightarrow \quad SX = 0.866 \times 100$$
$$= 86.6\,m$$

and   YX = TB = 50 m

Hence   SY = 86.6 m − 50 m

= 36.6 m

(d) Final bearing of tractor is $N\widehat{S}T$.

But $N\widehat{S}T = S\widehat{T}Y$ (alternate angles).

Using trigonometry in △STY

$$\tan S\widehat{T}Y = \frac{SY}{TY}$$

$$= \frac{36.6}{20}$$

so tan $S\widehat{T}Y = 1.83$

$S\widehat{T}Y = 61°$ (to nearest degree)

Final bearing is 061°.

Exercise 9.2

In this exercise give *all* distances and bearings correct to the nearest whole number. Do not use scale drawings.

1 A boat sails due south for 100 km and then due west for 50 km. What is the distance and bearing of the boat from its starting position?

2 A man walks due east for 6 km and then stops for a rest. He then walks on a bearing of 200°. How far has he walked on this bearing when he is exactly due south of his starting position? At this point how far south is he?

3 A bird (B) flies from its nest (O) on a bearing of 160° for 20 km. It then flies due north for 30 km. At this point how far is the bird from its nest and what is its bearing from the nest? (See Figure 9.14.)

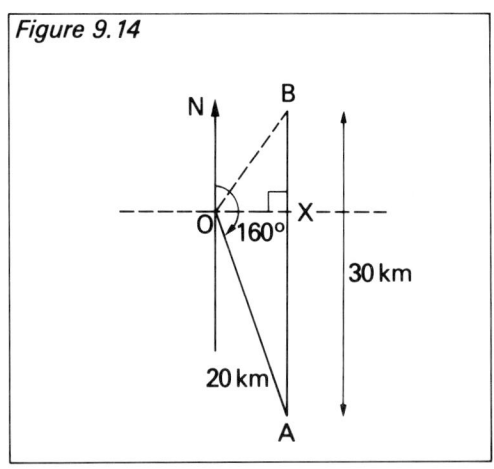

Figure 9.14

(Hint: Find OX and XA.)

4 A plane (P) flies due west from its airport (A) for 100 km before flying on a bearing of 280° for a further 200 km (see Figure 9.15).

Find   (a) how far the plane is due north
(b) how far the plane is due west   (c) the bearing of the plane from the airport.

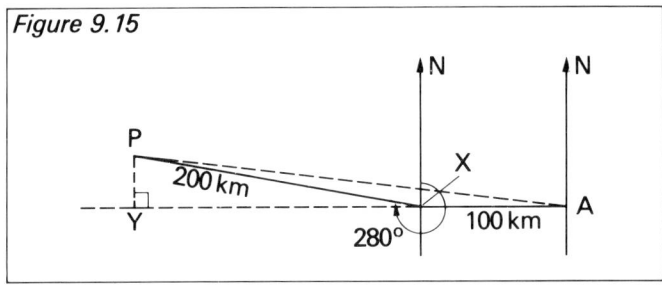

Figure 9.15

(Hint: Find XY and YP.)

5 A ship sails 50 km due east from a port, O, then 50 km due south. At this point it sails on a bearing of 205°. How far due south will it be when it reaches its destination which is on the north–south line through the original port, O? How far was the overall journey? (See Figure 9.16.)

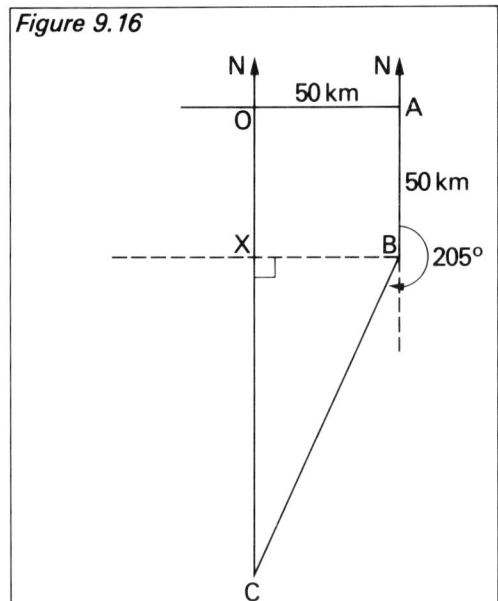

Figure 9.16

(Hint: Find $X\widehat{B}C$.)

6 A model boat sails off on a bearing of 060° for 10 m. It then sails due south for 10 m and then due east for a further 10 m (see Figure 9.17).

Find   (a) the overall distance travelled due east
(b) the overall distance travelled due south
(c) the bearing of the boat from its starting point.

(Hint: Find OP and XP.)

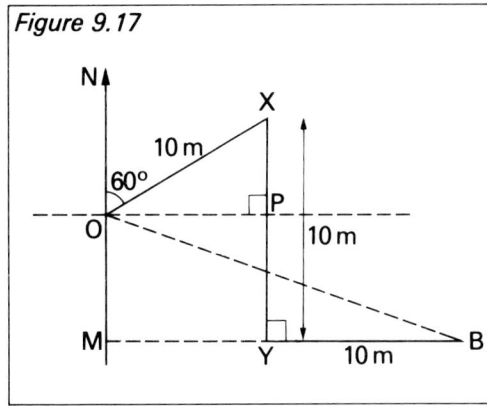

Figure 9.17

7 An insect crawls from a hole, H, on a bearing of 240° for 50 cm. It then crawls 80 cm due east (see Figure 9.18).

Find   (a) how far due south the insect crawled (b) how far due east from the hole the insect crawled   (c) the bearing of the insect from the hole.

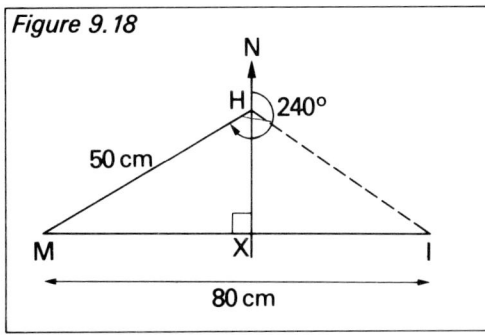

Figure 9.18

8 A lion runs from its den, D, on a bearing of 300° for 100 m. It then runs on a bearing of 030° until it is due north of its starting position (see Figure 9.19).

Find   (a) how far due north the lion is from its den   (b) the distance the lion will have travelled altogether if it returns directly to its den.

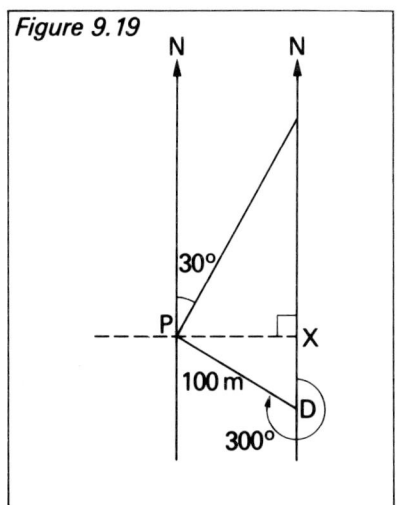

Figure 9.19

(Hint: Look carefully at all right-angled triangles.)

9 A tank drives in the desert for 50 km on a northerly bearing from O and then drives for a further 150 km on a bearing of 330°. The tank then drives due south for 160 km before coming to rest. What is the distance of the tank from its starting point and its bearing from this point? (See Figure 9.20.)

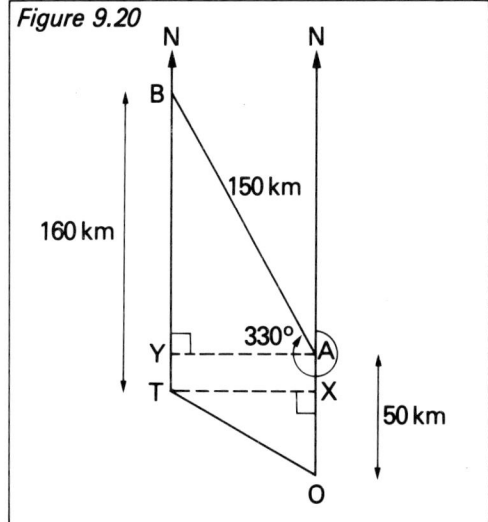

Figure 9.20

(Hint: Find AY and BY.)

10 A girl sets out from O and walks 3 km on a bearing of 040°. She then changes her bearing to 150° and walks for a further 8 km. How far is she from her starting point and what is her bearing from her starting point? (See Figure 9.21.)

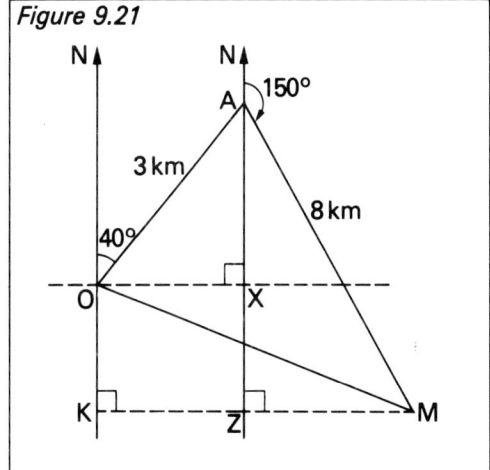

Figure 9.21

(Hint: Find OX, AX, AZ and MZ.)

10 Venn diagrams

Venn diagrams

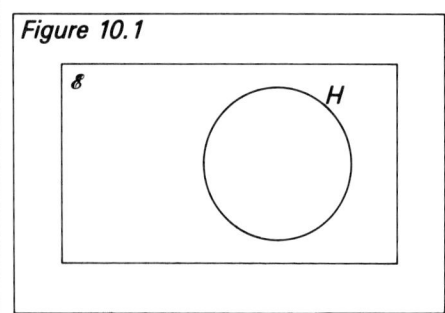

A VENN DIAGRAM IS THE DIAGRAMMATIC REPRESENTATION OF A UNIVERSAL SET AND ITS ASSOCIATED SUBSETS.

A UNIVERSAL SET IS REPRESENTED BY A RECTANGLE. SUBSETS OF A UNIVERSAL SET ARE USUALLY REPRESENTED BY CIRCLES DRAWN WITHIN THE RECTANGLE.

So if $\mathscr{E} = \{$mammals$\}$
and $H = \{$humans$\}$

representation of the universal set and this subset using a Venn diagram can be seen in Figure 10.1.

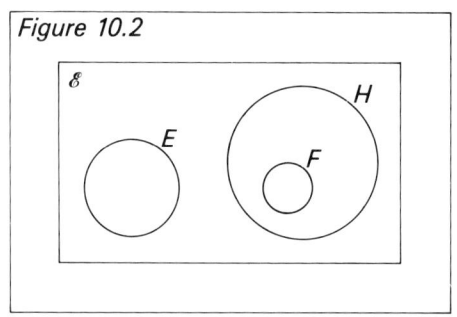

Figure 10.1

Further if $F = \{$human females$\}$
and $E = \{$elephants$\}$

the Venn diagram would now look like Figure 10.2.

Figure 10.2

Clearly, set F must lie completely inside set H since all human females belong exclusively to the set of humans. Since humans and elephants are independent species then the circles representing their sets must not intersect. Set H and set E are said to be *disjoint*.

Example 1

$\mathscr{E} = \{$humans$\}$
$A = \{$tall men$\}$
$B = \{$rugby players$\}$

Display this information in the form of a Venn diagram and shade in the region which represents the set $\{$tall men who play rugby$\}$.

Solution

See Figure 10.3.

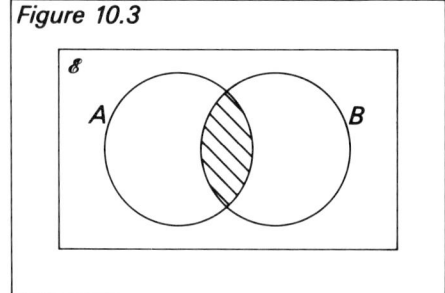

Figure 10.3

Clearly the intersection or overlap of the circles represents the set $\{$tall men who play rugby$\}$.

Shading regions

Example 2

Figure 10.4 shows a Venn diagram with the universal

set $\mathscr{E}$ and two subsets A and B. Shade the region represented by $A' \cap B$.

Figure 10.4

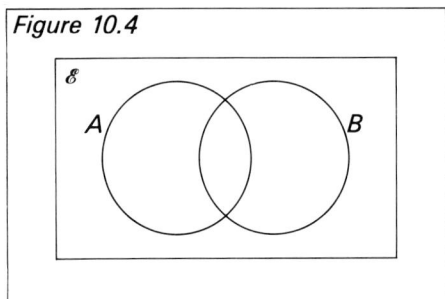

Solution

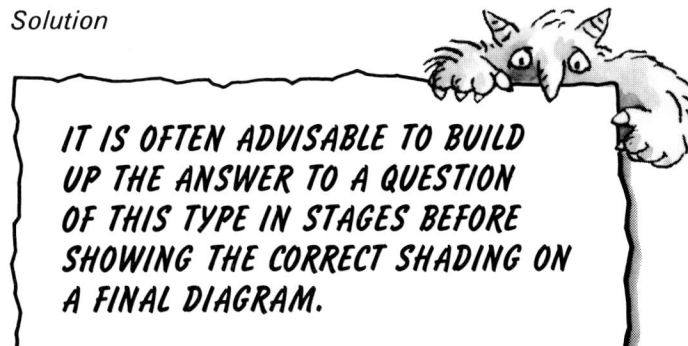

IT IS OFTEN ADVISABLE TO BUILD UP THE ANSWER TO A QUESTION OF THIS TYPE IN STAGES BEFORE SHOWING THE CORRECT SHADING ON A FINAL DIAGRAM.

(a) First shade in A' (see Figure 10.5).

Figure 10.5

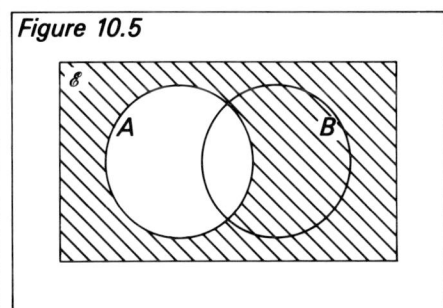

(b) On the same diagram shade in set B (see Figure 10.6).

Figure 10.6

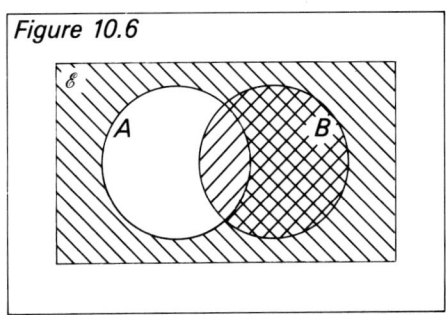

If the Venn diagram in Figure 10.6 is examined it will be seen that the region required is where the diagonal shading lines intersect. The final solution is shown in Figure 10.7.

Figure 10.7

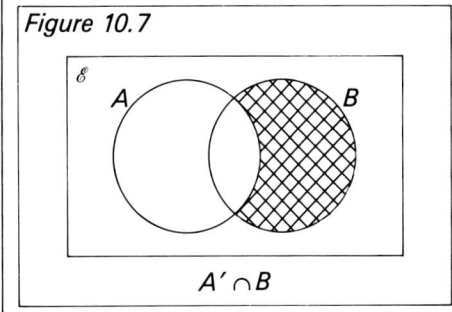

$A' \cap B$

Exercise 10.1

In questions 1 to 7 copy Figure 10.8 and, on a separate diagram for each part, shade in the region indicated.

Figure 10.8

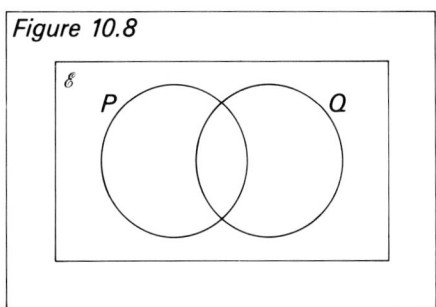

1 $P \cup Q$ **2** $P \cap Q$ **3** $P' \cup Q$ **4** $(P \cap Q)'$

5 $(P \cup Q)'$ **6** $P \cap Q'$ **7** $P' \cap Q'$

In questions 8 to 12 copy Figure 10.9 and, on a separate diagram for each part, shade in the region indicated.

Figure 10.9

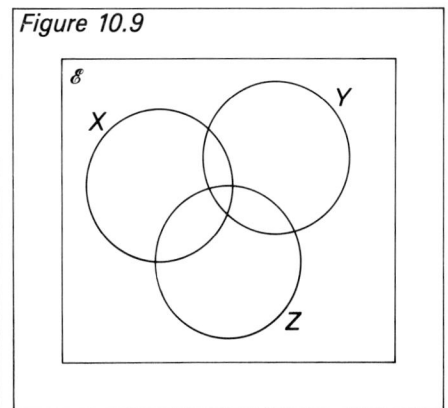

8 $(X \cup Y) \cap Z$ **9** $(X \cap Y) \cup Z$ **10** $X' \cap (Y \cup Z)$

11 $(X \cup Y \cup Z)'$ **12** $(X \cap Y) \cap Z'$

In questions 13 to 15 (over the page) copy Figure 10.10 and, on a separate diagram for each part, shade in the region indicated.

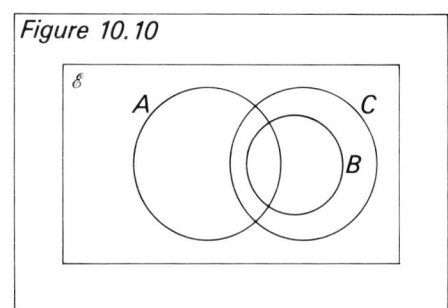

Figure 10.10

13 $A \cap B'$ **14** $(A \cap B)' \cap C$ **15** $(B' \cap C) \cap A'$

Problem solving using Venn diagrams

Example 3

In a survey, 100 people were asked if they watched BBC or ITV programmes. 63 people said that they watched BBC and 87 said that they watched ITV. Everyone watched TV.

How many watched   (a) BBC only   (b) ITV only (c) both stations?

Solution

Draw a Venn diagram. Let $B = \{$BBC viewers$\}$ and $I = \{$ITV viewers$\}$ (see Figure 10.11).

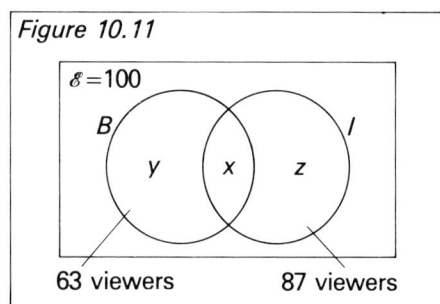

Figure 10.11

Let x be those people who watch both BBC and ITV.
Let y be those people who watch BBC only.
Let z be those people who watch ITV only.

BBC set
$$x + y = 63 \qquad (1)$$

ITV set
$$x + z = 87 \qquad (2)$$

Universal set
$$x + y + z = 100 \qquad (3)$$

Take equation (1) $x + y = 63$
$$y = 63 - x \qquad (4)$$
Take equation (2) $x + z = 87$
$$z = 87 - x \qquad (5)$$

Now substitute equation (4) and equation (5) into equation (3)

$$x + (63 - x) + (87 - x) = 100$$
$$150 - x = 100$$
$$x = 50$$

Substitute $x = 50$ into equation (1). This gives $y = 13$, which when substituted into equation (2) gives $z = 37$.

The solution is   (a) 13 watch BBC only (b) 37 watch ITV only   (c) 50 watch both stations.

Example 4

On a small estate of 35 houses a newsagent delivers the following publications:
(a) *The Times* – 20 copies
(b) *The Mirror* – 15 copies
(c) *The Beano* – 26 copies.
However the newsagent has some other information about his deliveries:
(d) 6 houses take all three publications
(e) 4 houses take *The Times* and *The Mirror* but not *The Beano*
(f) 3 houses take *The Mirror* and *The Beano* but not *The Times*
(g) 2 houses take *The Times* only.

Draw a Venn diagram to display this information and use the diagram to find the following:
(i) how many houses take *The Times* and *The Beano* but not *The Mirror*
(ii) how many houses take *The Beano* only
(iii) how many houses take both *The Times* and *The Beano*
(iv) how many houses take only one publication
(v) how many houses take no publication.

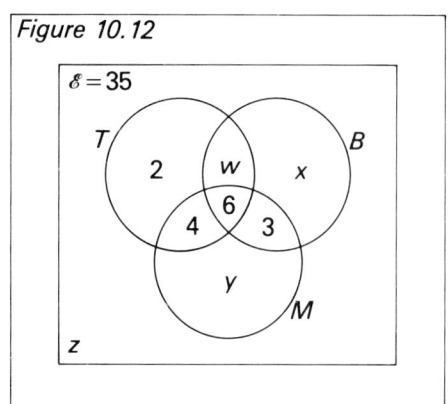

Figure 10.12

Solution

See Figure 10.12.

Let T = {houses taking *The Times*}
        B = {houses taking *The Beano*}
        M = {houses taking *The Mirror*}.

> THE INFORMATION GIVEN SHOULD BE PUT ON THE VENN DIAGRAM, STARTING FROM THE CENTRE ($T \cap B \cap M$) AND WORKING OUTWARDS.

i.e.   6 take all 3 publications ($T \cap B \cap M$)
    4 take *The Times* and *The Mirror* but not *The Beano* ($T \cap M \cap B'$)
    3 take *The Beano* and *The Mirror* but not *The Times* ($B \cap M \cap T'$)
    2 take *The Times* only ($(B \cup M)' \cap T$)

Now let w be those who take *The Times* and *The Beano* but not *The Mirror*
        x be those who take *The Beano* only
        y be those who take *The Mirror* only
        z be those who take no publications.

The Times set
$$w + 2 + 4 + 6 = 20$$
$$w + 12 = 20$$
$$w = 8$$

The Mirror set
$$y + 4 + 6 + 3 = 15$$
$$y + 13 = 15$$
$$y = 2$$

The Beano set
$$w + x + 6 + 3 = 26$$
but since $w = 8$,
$$8 + x + 6 + 3 = 26$$
$$x + 17 = 26$$
$$x = 9$$

Universal set
$$w + x + y + z + 2 + 4 + 6 + 3 = 35$$
but $w = 8$, $x = 9$, $y = 2$
so $8 + 9 + 2 + z + 2 + 4 + 6 + 3 = 35$
$$z + 34 = 35$$
$$z = 1$$

The questions can now be answered
(i)   houses which take *The Times* and *The Beano* but not *The Mirror* $(w) = 8$
(ii)   houses which take *The Beano* only $(x) = 9$
(iii) houses which take *The Times* and *The Beano* $(w + 6) = 14$
(iv)   houses which take only one publication $(2 + x + y) = 13$
(v)   houses which take no publication $(z) = 1$.

Exercise 10.2

1 In Figure 10.13, $\mathscr{E}$ = {integers x: $1 \leqslant x \leqslant 10$}.

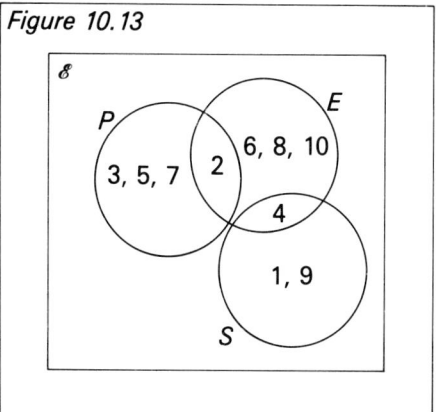

Figure 10.13

Write down the list of elements belonging to the following sets.   (a) P (b) E (c) S (d) $P \cap E$ (e) S' (f) $P \cap S$ (g) $(S \cup E)' \cap P$ (h) $(P \cup S) \cap E$.

Find   (i) $n(E')$ (j) $n(P \cap E \cap S)$.

2 In Figure 10.14 the dots represent the people who took part in a survey about the places they visited at a zoo. E, S and L are the sets of people who visit the elephant house, snake house and lion house respectively.

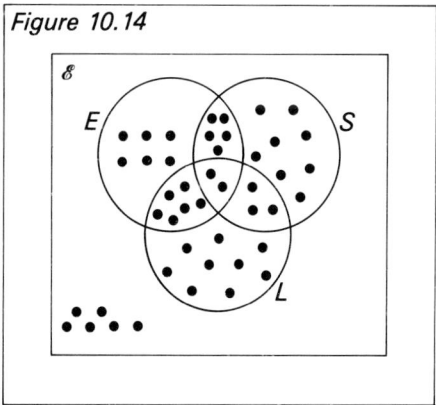

Figure 10.14

Find   (a) $n(\mathscr{E})$ (b) $n(E)$ (c) $n(S)$ (d) $n(L)$
(e) $n(E \cap L)$ (f) $n(E \cup S \cup L)$ (g) $n(E')$
(h) $n((S \cup L)')$ (i) $n((S \cup L \cup E)')$ (j) $n(S' \cap E)$.

What is the meaning of each of the following?
(k) $S \cap L \cap E$ (l) S' (m) $E \cap L$ (n) $E \cup L$
(o) $(E \cap L) \cup S$.

3 At a party there was beer and spirits for the
50 guests. 37 drank beer and 25 drank spirits. Draw
a Venn diagram to illustrate this information and use
it to find those who drank (a) beer only
(b) spirits only (c) both beer and spirits. (Assume
everyone has a drink.)

4 On a certain flight to the USA containing
120 tourists, 76 intended to visit New York,
40 intended to visit Washington and 13 intended to
visit both these cities. Draw a Venn diagram to
illustrate this information and find out how many
people do not intend to visit either New York or
Washington.

5 In Figure 10.15, $\mathscr{E} = \{$cars failing their MOT$\}$. Let
$S = \{$cars failing with faulty steering$\}$
$B = \{$cars failing with faulty brakes$\}$
$C = \{$cars failing with body corrosion$\}$.

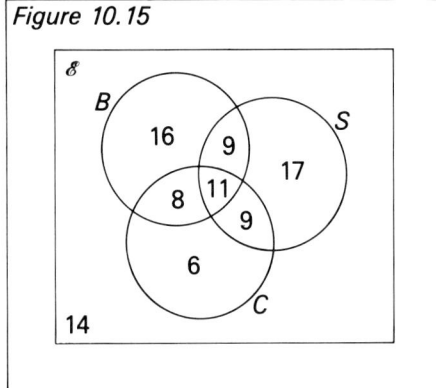

Figure 10.15

Find (a) $n(B)$ (b) $n(C)$ (c) $n(S)$ (d) $n(\mathscr{E})$
(e) $n(C')$ (f) $n(S \cap C)$ (g) $n(S \cap C \cap B)$
(h) $n((S \cup B)')$ (i) $n((S \cup C \cup B)')$ (j) $n(C' \cap B)$.
What is the meaning of the following?
(k) $(S \cup C \cup B)'$ (l) C' (m) $B \cap S$ (n) $B \cup S$.

6 25 pupils out of a class of 35 like maths while
27 like English. Only 3 do not like maths or English.
Represent this information on a Venn diagram and
find those who like (a) only maths
(b) only English (c) both maths and English.

7 200 third-year pupils were asked to choose their
science options for the fourth year with the
requirement that they must choose at least one
subject.

30 chose physics only
42 chose physics and chemistry but not biology
29 chose chemistry and biology but not physics

36 chose all three subjects
116 chose physics
125 chose chemistry
110 chose biology

Draw a Venn diagram to illustrate this information
and use the diagram to find the following
(a) how many chose physics and biology but not
chemistry (b) how many chose only one subject
(c) how many chose only biology (d) how many
chose physics and chemistry.

8 50 children are asked about their favourite foods.

23 like curry
24 like fish and chips
26 like spaghetti
8 like only curry
9 like curry and spaghetti but not fish and chips
7 like spaghetti and fish and chips but not curry
4 like all three foods

Draw a Venn diagram to illustrate the information
and then use it to find those who like (a) fish and
chips only (b) none of these foods (c) curry and
fish and chips (d) only one kind of food.

9 A group of 60 TV viewers were asked which type of
programme they preferred.

39 liked horror films
16 liked plays
28 liked westerns
11 liked horror films and westerns but not plays
5 liked horror films and plays but not westerns
6 liked all three types of programme
3 liked only plays

Draw a Venn diagram to display this information
and use it to find those viewers who liked
(a) westerns only (b) horror films only
(c) one type of programme only
(d) westerns and plays (e) only two of the three
types of programme (f) none of these programmes.

10 At a camping/caravan exhibition a survey of
80 people showed that 19 had not been on a
camping or caravan holiday. 53 had been camping
and 45 had been on a caravan holiday. Draw a Venn
diagram to display this information and find how
many people had been on (a) both types of
holiday (b) only one type of holiday.

11 Out of 100 young craftsmen, 38 had been on a
goldsmiths' course, 32 on a silversmiths' course
and 13 on both these courses. Draw a Venn diagram
to illustrate this information and find (a) how many
attended only one course (b) how many had not
been on either course.

12 On 'games evening' 80 customers at a local pub enjoy playing the following games.

53 like darts
53 like dominoes
36 like draughts
 7 like darts only
12 like darts and draughts but not dominoes
21 like darts and dominoes but not draughts
everyone likes something

Draw a Venn diagram to display all the information and use it to find how many like
(a) dominoes only   (b) all three games
(c) only one game   (d) dominoes and draughts
(e) exactly two games.

13 In a survey of 100 people, 8 did not like chocolates or toffee, but 78 liked chocolates and 71 liked toffee. Draw a Venn diagram to illustrate this information and find those who   (a) like chocolate and toffee   (b) only like one type of sweet.

14 50 sixth-formers choose their 'A' level courses as follows.

24 take maths
24 take chemistry
19 take physics
 3 take all three subjects
 6 take maths and physics
10 take maths and chemistry
 7 take physics and chemistry
the remainder take biology

Draw a Venn diagram to show this information and use it to answer the following questions. How many choose   (a) biology only
(b) maths and chemistry but not physics
(c) only one subject   (d) at least two subjects?

15 Out of 100 boys

62 play football
57 play rugby
31 play football and rugby
38 play tennis
27 play football and tennis
18 play rugby and tennis
13 play all three sports

Draw a Venn diagram to illustrate this information and use the diagram to answer the following questions. How many boys   (a) play none of these sports   (b) play at least 2 sports   (c) play only one sport?

11 Mensuration

Area and perimeter

(a) Triangle

Area $= \frac{1}{2} \times$ base $\times$ perpendicular height

or $A = \frac{1}{2}bh$ (see Figure 11.1)

Figure 11.1
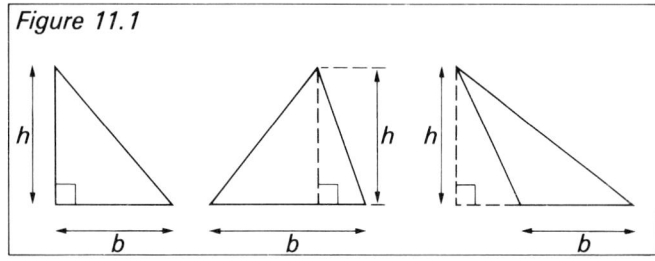

(b) Rectangle

Area $=$ length $\times$ width

or $A = LW$ (see Figure 11.2)

Figure 11.2
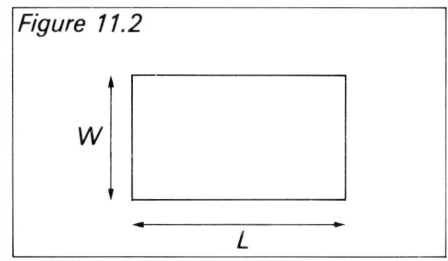

Perimeter $= 2 \times$ sum of length and width

or $P = 2(L + W)$

(c) Parallelogram

Area $=$ base $\times$ perpendicular height

or $A = bh$ (see Figure 11.3)

Figure 11.3
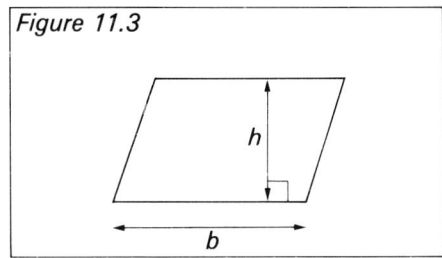

(d) Trapezium

Area $= \frac{1}{2} \times$ sum of parallel sides $\times$ perpendicular height

or $A = \frac{1}{2}(a + b)h$ (see Figure 11.4)

Figure 11.4
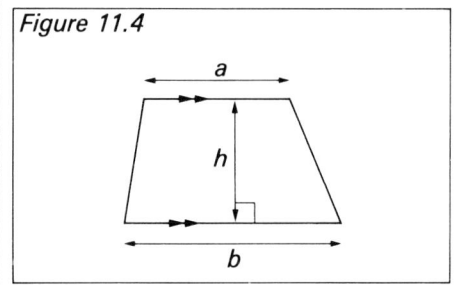

(e) Circle

Area $= \pi \times$ radius $\times$ radius

or $A = \pi r^2$ (see Figure 11.5)

π (PRONOUNCED 'PI') IS A LETTER OF THE GREEK ALPHABET. IN MATHEMATICS IT REPRESENTS THE NUMBER WITH THE VALUE $3\frac{1}{7}$, OR $\frac{22}{7}$, OR 3.14 (TO 2 D.P.).

Figure 11.5
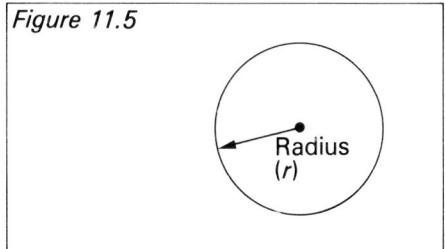

Circumference $= 2 \times \pi \times$ radius

or $C = 2\pi r$

(f) Sector

Area $= \dfrac{a}{360°} \times$ area of a circle

where a is the angle subtended at the centre of the circle by the arc and two radii bounding the sector (see Figure 11.6).

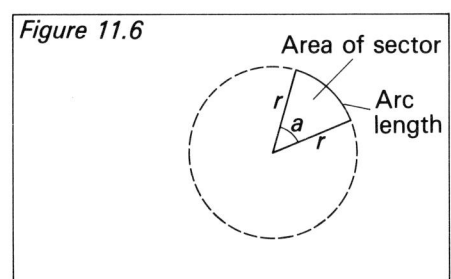

Figure 11.6

so
$$A = \frac{a}{360°} \times \pi r^2$$

Length of an arc $= \dfrac{a}{360°} \times$ circumference of a circle

or
$$L = \frac{a}{360°} \times 2\pi r$$

$$= \frac{a}{180°} \times \pi r$$

(g) Areas of similar shapes
In Figure 11.7, shapes A_1 and A_2 are similar.

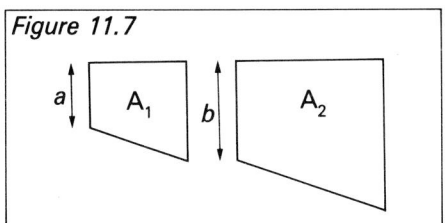

Figure 11.7

THE RATIO BY WHICH THE LENGTH OF ONE SHAPE IS LARGER THAN THE CORRESPONDING LENGTH OF A SIMILAR SHAPE IS CALLED THE <u>SCALE FACTOR</u>.

So   scale factor $= \dfrac{b}{a}$

and   area of $A_2 = \left(\dfrac{b}{a}\right)^2 \times$ area of A_1

Example 1

In Figure 11.8, the two shapes are similar. If the area of the smaller shape is 245 cm², find the area of the larger shape.

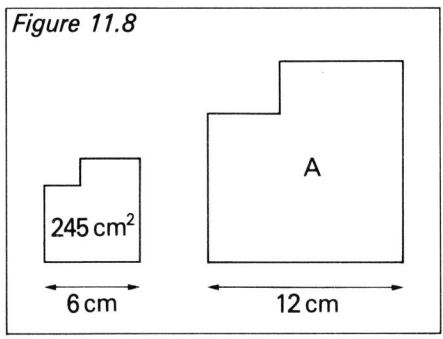

Figure 11.8

Solution

Scale factor of enlargement $= \dfrac{12}{6} = 2$

and area of A $=$ (scale factor)$^2 \times$ area of smaller shape

$$= 2^2 \times 245$$
$$= 4 \times 245$$
area of A $= 980\,\text{cm}^2$

Example 2

Find the area of the irregular pentagon ABCDE in Figure 11.9 which consists of triangle ABE and parallelogram BCDE.

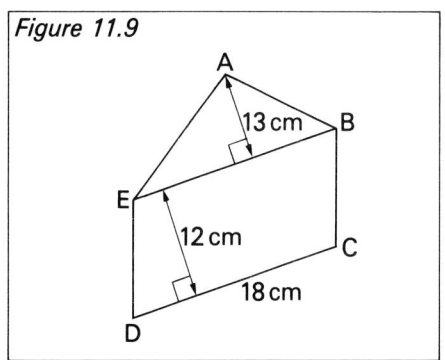

Figure 11.9

Solution

Area of   Area of   Area of
pentagon   =   triangle + parallelogram
ABCDE   ABE   BCDE

Area of triangle ABE $= \frac{1}{2}bh$
$$= \frac{1}{2} \times 18 \times 13$$
$$= 9 \times 13$$
$$= 117\,\text{cm}^2$$

Area of parallelogram BCDE = bh
$$= 18 \times 12$$
$$= 216\,\text{cm}^2$$

Hence   area of pentagon = $117\,\text{cm}^2 + 216\,\text{cm}^2$
$$= 333\,\text{cm}^2$$

Example 3

Find the area and perimeter of the lawn whose shape is shown in Figure 11.10. Let $\pi = \frac{22}{7}$.

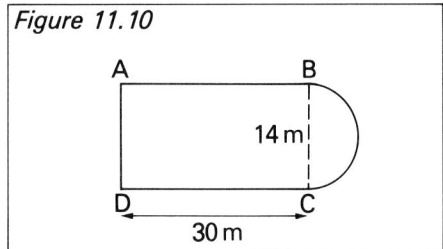

Figure 11.10

Solution

(a) $\dfrac{\text{Area of}}{\text{lawn}} = \dfrac{\text{area of}}{\text{rectangle}} + \dfrac{\text{area of}}{\text{semi-circle}}$

Area of rectangle = LW
$$= 30 \times 14$$
$$= 420\,\text{m}^2$$

Area of semi-circle = $\frac{1}{2}\pi r^2$

$$= \frac{1}{2} \times \frac{\overset{11}{\cancel{22}}}{\underset{1}{\cancel{7}}} \times 7 \times \overset{1}{\cancel{7}}$$

$$= 77\,\text{m}^2$$

Area of lawn = $420\,\text{m}^2 + 77\,\text{m}^2$
$$= 497\,\text{m}^2$$

(b) Perimeter of lawn = AB + arc BC + CD + DA.

AB = CD = 30 m,   DA = 14 m

Arc BC = $\frac{1}{2} \times$ circumference of circle

$$= \frac{1}{2} \times \frac{\overset{11}{\cancel{22}}}{\underset{1}{\cancel{7}}} \times \overset{2}{\cancel{14}}$$

$$= 22\,\text{m}$$

Perimeter = $(30 + 22 + 30 + 14)\,\text{m}$
$$= 96\,\text{m}$$

Example 4

Find the area and perimeter of a sector of a circle which subtends an angle at the centre of 70°. The radius of the circle is 6 cm. Let $\pi = \frac{22}{7}$.

Solution

See Figure 11.11.

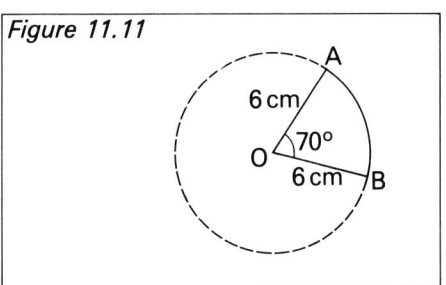

Figure 11.11

(a) Area = $\dfrac{a}{360°} \times \pi r^2$

$$= \frac{70}{360} \times \frac{22}{7} \times 6 \times 6$$

$$= \frac{70 \times 22 \times 6 \times 6}{360 \times 7}$$

Simplify the fraction where possible

Area = $\dfrac{\overset{1}{\cancel{70}} \times 22 \times 6 \times 6}{\underset{36}{\cancel{360}} \times \cancel{7}}$ (divide top and bottom first by 10 then by 7)

$$= \frac{22 \times \overset{1}{\cancel{6}} \times \overset{1}{\cancel{6}}}{\underset{1}{\cancel{\underset{6}{36}}}}$$ (divide top and bottom first by 6 then by 6 again)

area = $22\,\text{cm}^2$

(b) Perimeter = OA + arc AB + BO.

OA = BO = 6 cm

Arc AB = $\dfrac{a\pi r}{180°}$

$$= \frac{70}{180} \times \frac{22}{7} \times 6$$

$$= \frac{70 \times 22 \times 6}{180 \times 7}$$

Simplify the fraction where possible

Arc AB = $\dfrac{\overset{1}{\cancel{70}} \times 22 \times 6}{\underset{18}{\cancel{180}} \times \underset{1}{\cancel{7}}}$ (divide top and bottom first by 10 then by 7)

$$= \frac{22 \times \overset{1}{\cancel{6}}}{\underset{3}{\cancel{18}}}$$ (divide top and bottom by 6)

$$= \frac{22}{3}$$

arc AB = $7\frac{1}{3}\,\text{cm}$

Perimeter = $(6 + 7\frac{1}{3} + 6)\,\text{cm}$
$$= 19\frac{1}{3}\,\text{cm}$$

Exercise 11.1

1 Find the area and perimeter of the quadrilateral ABCD in Figure 11.12.

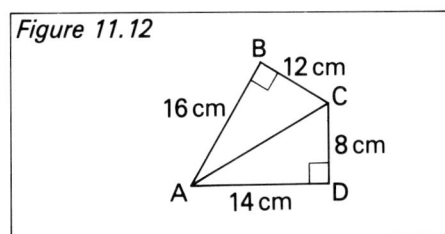

Figure 11.12

2 Find the area of the quadrilateral PQRS in Figure 11.13.

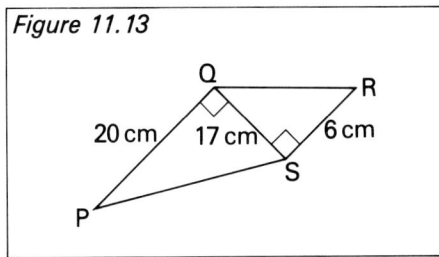

Figure 11.13

3 Find the area of the trapezium WXYZ in Figure 11.14.

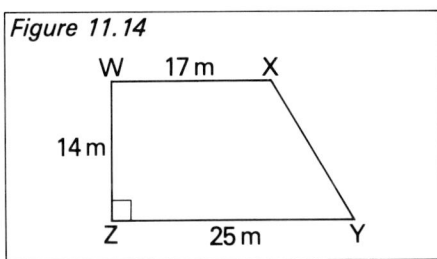

Figure 11.14

4 In Figure 11.15 triangle ABC is similar to triangle DEF. If triangle ABC has an area of 9 cm², find the area of triangle DEF.

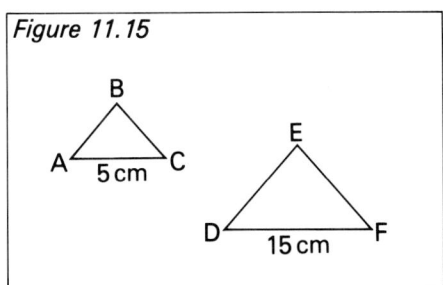

Figure 11.15

5 Find the area and perimeter of the shape in Figure 11.16. Let $\pi = \frac{22}{7}$.

6 Figure 11.17 shows a place mat. Find its area and perimeter. Let $\pi = \frac{22}{7}$.

7 Figure 11.18 shows the plan of a ground floor room. Find its area and perimeter.

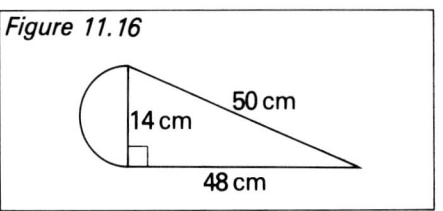

Figure 11.16

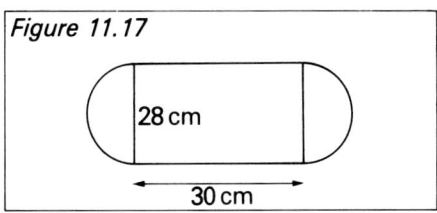

Figure 11.17

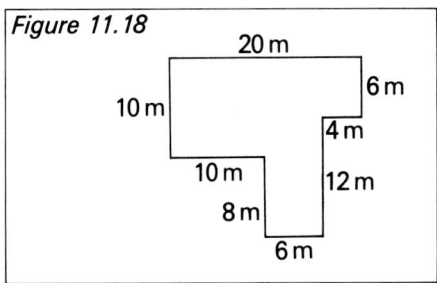

Figure 11.18

8 In Figure 11.19 the two shapes are similar. If the smaller shape has an area of 44 cm², find the area of the larger shape.

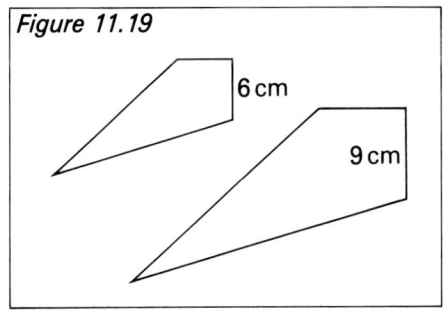

Figure 11.19

9 Find the area and perimeter of the letter 'E' shown in Figure 11.20.

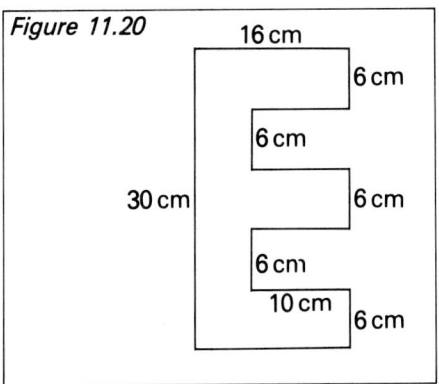

Figure 11.20

10 Two rectangles are similar. The length of the smaller one is 9 cm and the length of the larger one is 27 cm. Find the area of the larger rectangle if the area of the smaller one is 76 cm².

11 The two shapes in Figure 11.21 are similar. Find the area of the smaller shape given that the area of the larger shape is 544 cm².

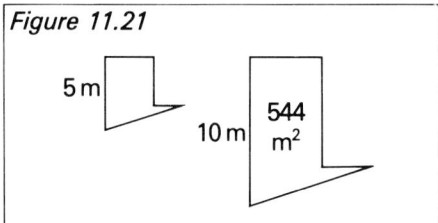

Figure 11.21

5 m

10 m

544 m²

12 The shape in Figure 11.22 consists of a rectangle, parallelogram and semi-circle. Find its total area. Let $\pi = \frac{22}{7}$.

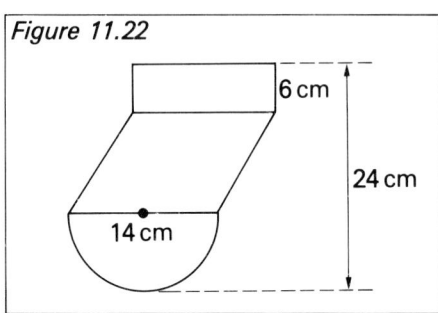

Figure 11.22

6 cm

24 cm

14 cm

13 Find the area and perimeter of the shape in Figure 11.23. Let $\pi = \frac{22}{7}$.

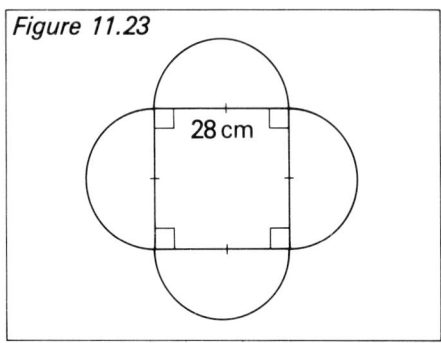

Figure 11.23

28 cm

14 Find the area and perimeter of the shape in Figure 11.24.

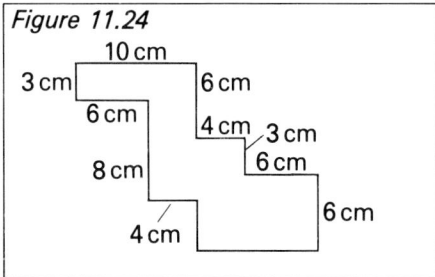

Figure 11.24

10 cm

3 cm

6 cm

6 cm

4 cm   3 cm

8 cm

6 cm

4 cm

6 cm

15 Figure 11.25 shows the plan of a circular ornamental pond surrounded by a circular flower bed. Find the area of the flower bed. Let $\pi = \frac{22}{7}$.

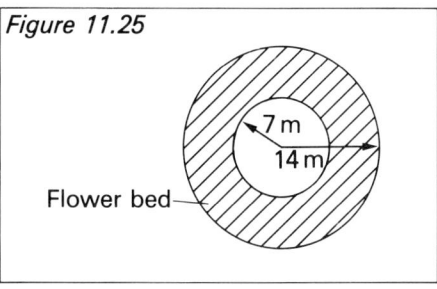

Figure 11.25

7 m

14 m

Flower bed

16 Find the shaded area in Figure 11.26. Let $\pi = 3.14$.

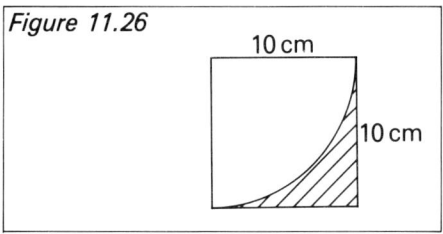

Figure 11.26

10 cm

10 cm

17 Two similar shapes have corresponding lengths of 3 cm and 18 cm. If the area of the smaller shape is 15 cm², find the area of the larger shape.

18 Find the area and perimeter of the shaded shape in Figure 11.27. Let $\pi = \frac{22}{7}$.

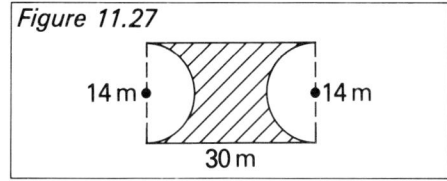

Figure 11.27

14 m

14 m

30 m

19 Find the area and perimeter of the shaded region inside the rhombus in Figure 11.28.

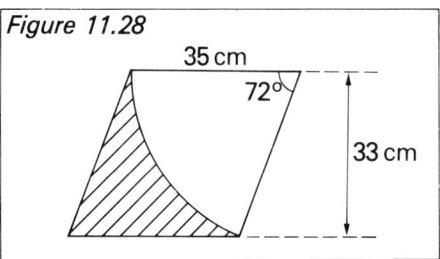

Figure 11.28

35 cm

72°

33 cm

20 Figure 11.29 consists of two semi-circles attached to a trapezium. Find the total area. Let $\pi = \frac{22}{7}$.

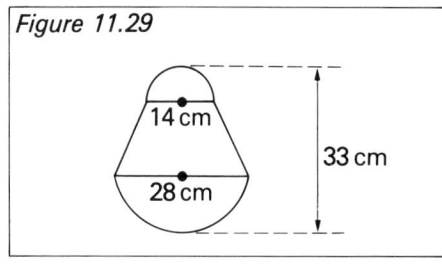

Figure 11.29

21 Find the area and perimeter of the irregular hexagon ABCDEF shown in Figure 11.30.

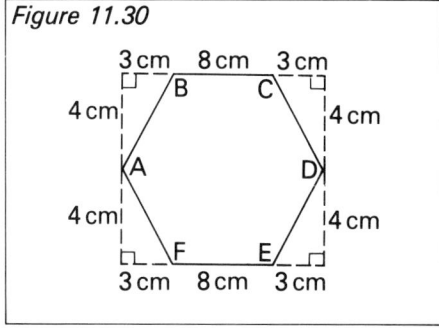

Figure 11.30

22 Find the area of the irregular pentagon ABCDE shown in Figure 11.31.

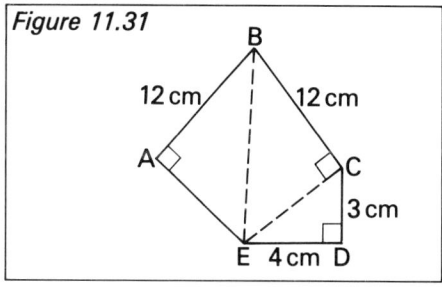

Figure 11.31

23 Find the area and perimeter of the quadrilateral PQRS in Figure 11.32, given that PQ = RS.

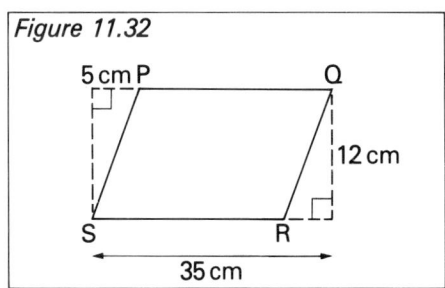

Figure 11.32

24 Find the shaded area in the trapezium shown in Figure 11.33, when the semi-circle has been removed. Let $\pi = \frac{22}{7}$.

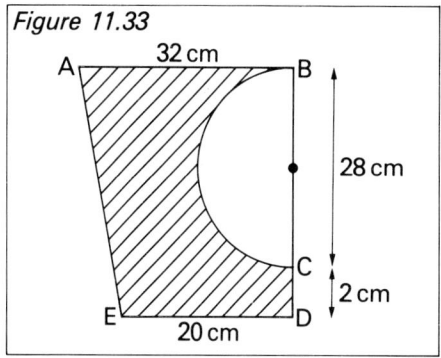

Figure 11.33

25 Calculate the total area of the fan blades shown in Figure 11.34 if the angle subtended by each blade is 63° and the radius is 10 cm. Let $\pi = \frac{22}{7}$.

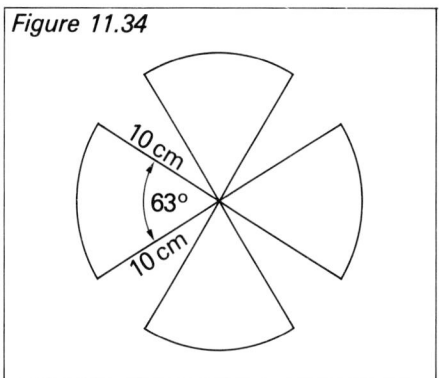

Figure 11.34

Volume

(a) Cuboid

Volume = length × width × height
or $V = LWH$ (see Figure 11.35)

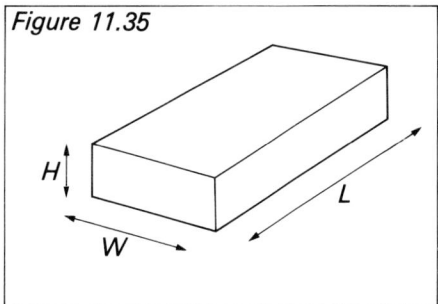

Figure 11.35

(b) Prism

Volume = cross-sectional area × length

or $V = AL$ (see Figure 11.36)

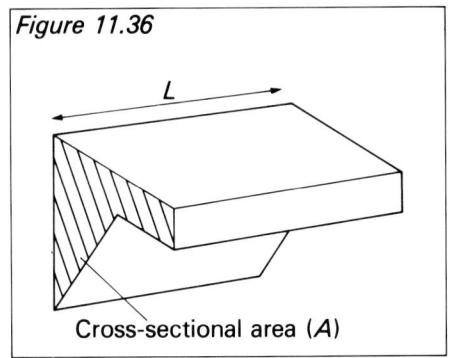

Figure 11.36

L

Cross-sectional area (A)

(c) Cylinder

Volume = area of end × height

or $V = \pi r^2 h$ (see Figure 11.37)

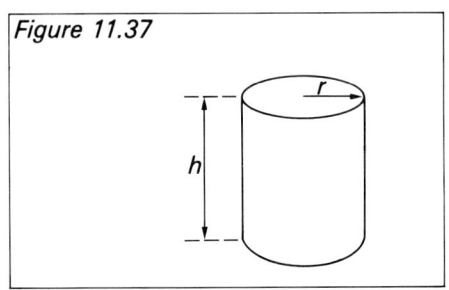

Figure 11.37

r

h

(d) Cone

Volume = $\frac{1}{3}$ × area of base × height

or $V = \frac{1}{3}\pi r^2 h$ (see Figure 11.38)

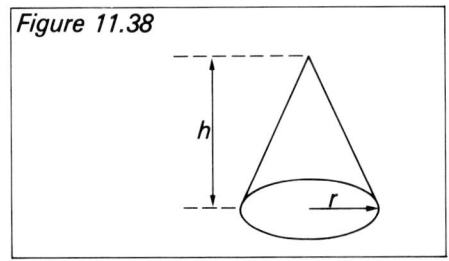

Figure 11.38

h

r

(e) Pyramid

Volume = $\frac{1}{3}$ × area of base × height

or $V = \frac{1}{3}Ah$

Figure 11.39 shows two examples of pyramids. One has a triangular base and the other a rectangular base.

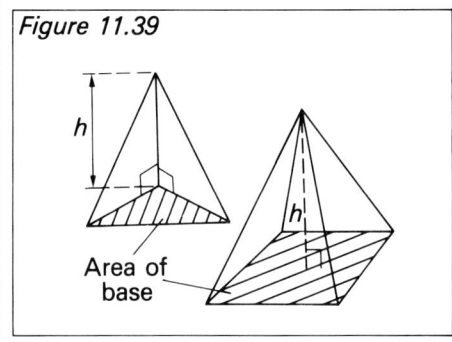

Figure 11.39

h

Area of base

h

(f) Sphere

Volume = $\frac{4}{3}$ × π × radius × radius × radius

or $V = \frac{4}{3}\pi r^3$ (see Figure 11.40)

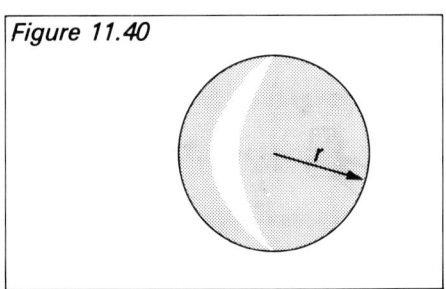

Figure 11.40

r

(g) Volumes of similar objects

In Figure 11.41, the objects V_1 and V_2 are similar.

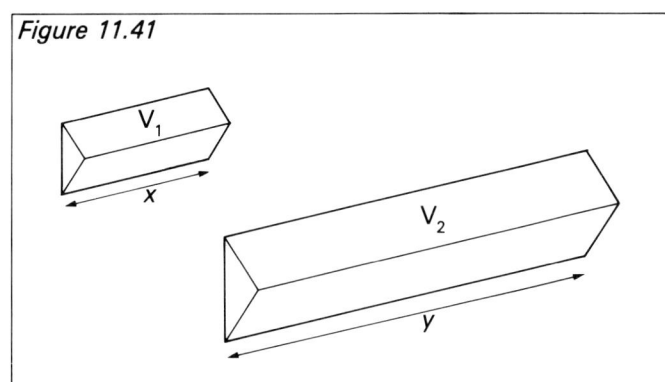

Figure 11.41

V_1

x

V_2

y

The ratio by which the length of a side of one object is larger than the length of a corresponding side of a similar object is called the scale factor.

Hence   scale factor $= \dfrac{y}{x}$

and   volume of $V_2 = \left(\dfrac{y}{x}\right)^3 \times$ volume of V_1

Example 5

Find the volume of the tetrahedron (triangular-based pyramid) shown in Figure 11.42.

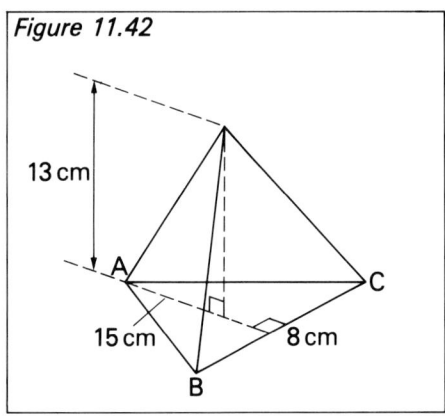

Figure 11.42

13 cm

A

15 cm   8 cm

B

C

Solution

Volume $= \frac{1}{3} \times$ base area $\times$ height
$= \frac{1}{3}Ah$

Base area $=$ area of triangle ABC
$= \frac{1}{2} \times 8 \times 15$
$= 60\,\text{cm}^2$

Volume $= \frac{1}{3} \times 60 \times 13$
$= 20 \times 13$
$= 260\,\text{cm}^3$

Example 6

A prism has a length of 4 cm and a volume of 232 cm³.
Find the volume of a similar prism having a length
of 6 cm.

Solution

Find the scale factor by which the length of one prism is
larger than the other.

Length of larger prism $= 6\,\text{cm}$
Length of smaller prism $= 4\,\text{cm}$

Length of larger prism $= \dfrac{6}{4} \times$ length of smaller prism

$= \dfrac{3}{2} \times$ length of smaller prism

The scale factor is $\frac{3}{2}$.

Volume of
larger prism $=$ (scale factor)³ $\times$ Volume of
smaller prism

$= \left(\dfrac{3}{2}\right)^3 \times 232$

$= \dfrac{27}{\underset{1}{\cancel{8}}} \times \overset{29}{\cancel{232}}$

$= 27 \times 29$
$= 783\,\text{cm}^3$

Example 7

Figure 11.43 shows a spacer in a special engine which
is made from a cuboid of metal with a hole drilled down
the centre. Find the volume and mass of metal
(to nearest kg) if the density of the metal is 5.8 g/cm³.
Let $\pi = \frac{22}{7}$.

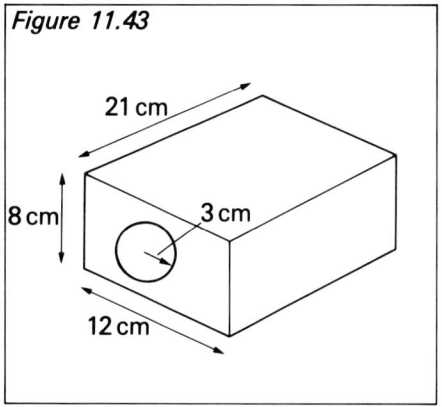

Figure 11.43

21 cm

8 cm   3 cm

12 cm

Solution

(a) Volume of metal $=$ volume of cuboid $-$ volume of hole (cylinder)

Volume of cuboid $= LWH$
$= 21 \times 12 \times 8$
$= 2016\,\text{cm}^3$

Volume of cylinder $= \pi r^2 h$

$= \dfrac{22}{\underset{1}{\cancel{7}}} \times 3 \times 3 \times \overset{3}{\cancel{21}}$

$= 594\,\text{cm}^3$

Volume of metal $= (2016 - 594)\,\text{cm}^3$
$= 1422\,\text{cm}^3$

(b) Mass of metal $=$ volume of metal $\times$ density
$= (1422 \times 5.8)\,\text{g}$
$= 8247.6\,\text{g}$
$= 8.2476\,\text{kg}$
$= 8\,\text{kg}$ (to nearest kg)

Exercise 11.2

1 Find the volumes of the cuboids whose dimensions
are given in the table below.

Question	(a)	(b)	(c)	(d)	(e)	(f)
Length (L)	20 cm	8 m	3 cm	4 m	730 cm	120 cm
Width (W)	5 cm	3.5 m	8 mm	1.2 m	4.5 m	0.9 m
Height (H)	3 cm	2.5 m	5 mm	0.8 m	2 m	0.8 m

2 Find the volumes of the cylinders whose dimensions are given in the table below. Let $\pi = \frac{22}{7}$ in each case.

Question	(a)	(b)	(c)	(d)	(e)	(f)
Radius (r)	3 cm	6 cm	7 m	21 mm	$3\frac{1}{2}$ cm	$2\frac{1}{2}$ m
Height (h)	7 cm	14 cm	15 m	12 mm	16 cm	28 m

3 Find the volume of each prism in Figure 11.44 giving your answers in cm³.

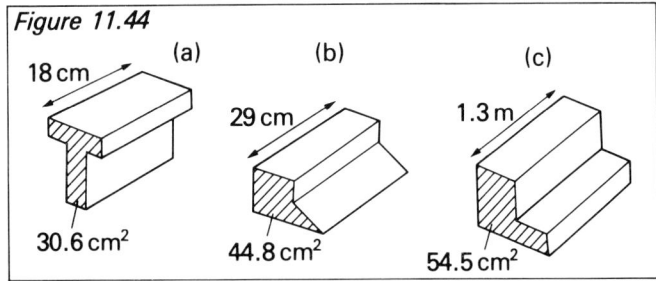

Figure 11.44

(a) 18 cm 30.6 cm²

(b) 29 cm 44.8 cm²

(c) 1.3 m 54.5 cm²

4 Find the volume of a cone which has a base radius of 9 cm and a height of 14 cm. Let $\pi = \frac{22}{7}$.

5 A pyramid has a square base with a side of 15 cm. What is its volume if the height is 12 cm?

6 A sphere has a radius of 6 cm. What is its volume to the nearest cm³? (Let $\pi = 3.14$.)

7 Find the volume of the pyramid shown in Figure 11.45.

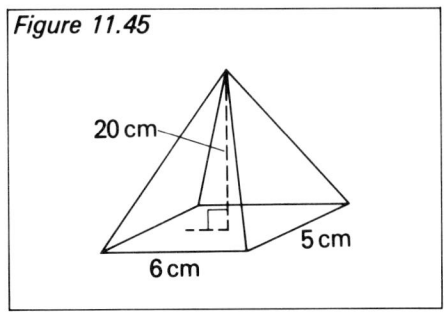

Figure 11.45

20 cm
5 cm
6 cm

8 Find the volume of the pyramid shown in Figure 11.46.

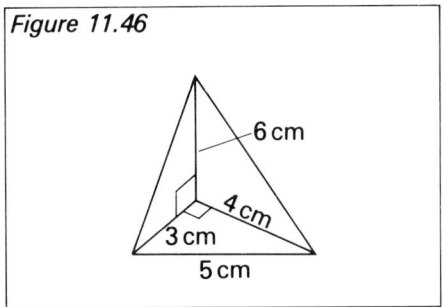

Figure 11.46

6 cm
4 cm
3 cm
5 cm

9 A cylinder of metal has a base radius of 6 cm and a height of 25 cm. Using $\pi = 3.14$, find the volume of the cylinder to the nearest cm³. Using the value thus obtained, find the mass of the cylinder if 1 cm³ of the metal weighs 2.5 g.

10 Find the volume of the 'I' section girder shown in Figure 11.47 giving your answer in cm³.

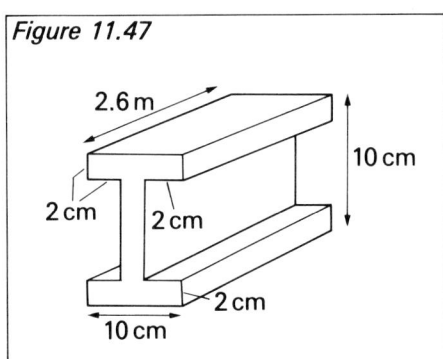

Figure 11.47

2.6 m
10 cm
2 cm
2 cm
2 cm
10 cm

11 A sphere has a volume of 100 cm³. What would be the volume of a sphere whose radius is twice as large?

12 A water tank has a square base with sides of 1.2 m and a height of 2 m. What is the volume in m³? Given that one litre of water weighs 1 kg and that 1 m³ contains 1000 litres, find the mass of water in the tank when it is half full. (Give your answer in tonne.)

13 A hexagonal tank has a volume of 1200 m³. What would be the volume of a similar tank having a height which was three times as large?

14 Two cuboids are made of different metals. One is made of molygold and has a length of 16 cm, width of 5.5 cm and a height of 8 cm. Molygold weighs 5.8 g per cm³. The other cuboid is made of aluminigum and has a length of 21 cm, width of 8.8 cm and a height of 10 cm. This metal weighs 2.2 g per cm³. Find the mass of each cuboid and state by how much one is heavier than the other.

15 A bar of gold in the shape of a cuboid is to be melted down and recast into small trinkets in the shape of square-based pyramids. The dimensions of

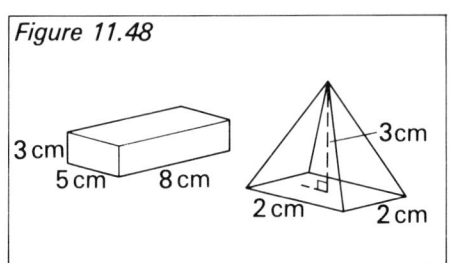

Figure 11.48

3 cm
5 cm
8 cm
3 cm
2 cm
2 cm

the cuboid and pyramid are given in Figure 11.48. Find out how many pyramids can be cast from the bar of gold.

16 The peg shown in Figure 11.49 is made of a cylinder of wood with a cuboid head. What is the volume of wood used? If the density of the wood is 0.9 g/cm³, find the mass of wood used in grams. Let $\pi = \frac{22}{7}$.

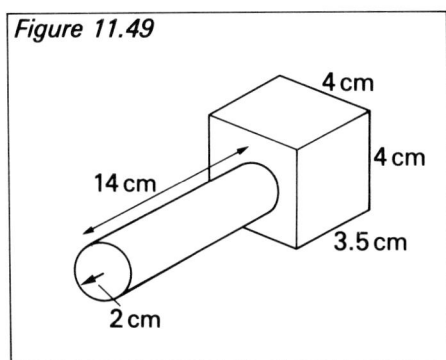

Figure 11.49

4 cm
4 cm
14 cm
3.5 cm
2 cm

17 A small cone has a base radius of 5 cm and a volume of 200 cm³. What is the base radius of a similar cone which has a volume of 1600 cm³?

18 A set of wheels and axle are made from solid steel and consist of two halves of a sphere connected by a cylinder (see Figure 11.50). The radius of the sphere is 7 cm and the radius of the cylinder is 3 cm. The length of the cylinder is 35 cm. Find the volume of steel required correct to the nearest cm³.
Let $\pi = \frac{22}{7}$.

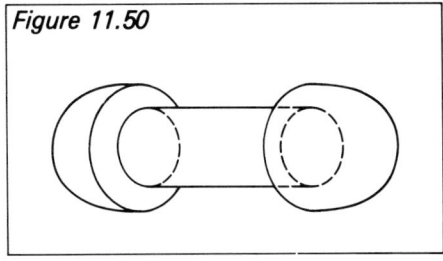

Figure 11.50

19 A cuboid of wood has seven holes drilled through it (see Figure 11.51). The radius of each hole is 1 cm. Find the volume of wood remaining when the seven holes have been drilled through. Let $\pi = \frac{22}{7}$.

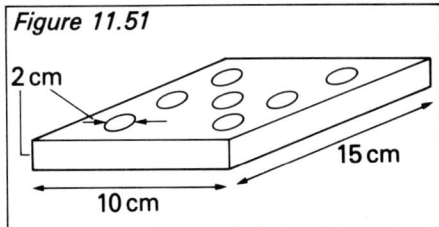

Figure 11.51

2 cm
15 cm
10 cm

20 What would be the volume of a prism which has a length of 24 cm when a similar shaped prism with a volume of 50 cm³ has a length of 8 cm?

21 A bar of silver has a cross-section in the shape of a trapezium (see Figure 11.52). Find its volume in cm³ and its mass to the nearest kg if 1 cm³ of silver weighs 10.5 g.

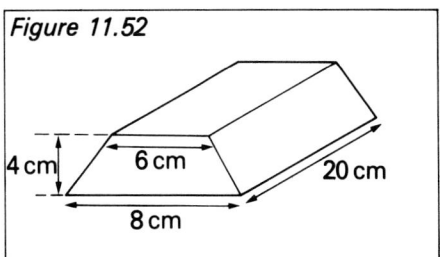

Figure 11.52

4 cm
6 cm
20 cm
8 cm

22 A special pipe is made by drilling a hole through the centre of a cylinder of solid steel (see Figure 11.53). The radius of the solid cylinder is 6 cm and the radius of the hole is 4 cm. The length of the pipe is 49 cm. Calculate, in cm³, the volume of steel drilled out of the pipe to make the hole. (Let $\pi = \frac{22}{7}$.) What is the volume of material remaining? What is the mass of the pipe to the nearest kg if 1 cm³ of metal weighs 7.9 g?

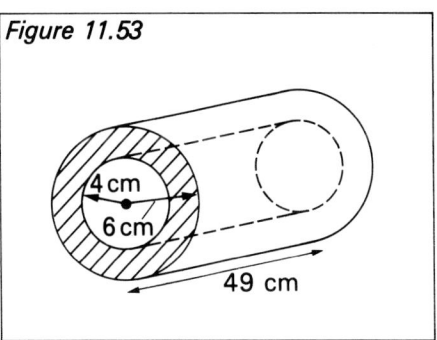

Figure 11.53

4 cm
6 cm
49 cm

23 A length of plastic drainpipe is shown in Figure 11.54. The radius of the inside surface is $2\frac{1}{2}$ cm and the outside surface is 3 cm. What is the volume of plastic in a 1.4 m length? (Give your answer in cm³ and let $\pi = \frac{22}{7}$.)

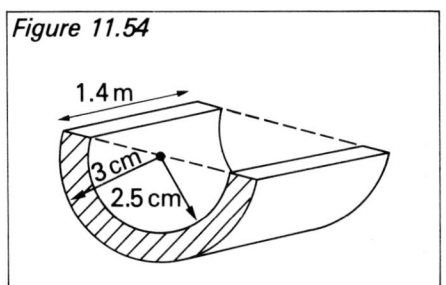

Figure 11.54

1.4 m
3 cm
2.5 cm

24 A cuboid having a volume of 5600 cm³ has a width of 12 cm. What would be the width of a similar cuboid having a volume of 700 cm³?

25 A railway tunnel is cut through a hillside in the shape of a semi-circle (see Figure 11.55). The diameter of the semi-circle is 14 m and the length of the tunnel is 120 m. Find the total volume of material removed. If $\frac{1}{3}$ of this is rock, $\frac{1}{8}$ shale, $\frac{7}{24}$ sandstone and the rest clay, then find what volume of each constituent has been removed.

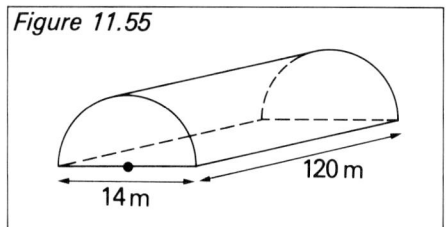

Figure 11.55

14 m

120 m

12 Statistics 2

Tally charts, frequency tables and class intervals

Example 1

The number of goals scored by the twenty-two teams in the first division of the football league on a Saturday afternoon in December was

0, 1, 2, 2, 3, 0, 2, 0, 4, 1, 0,
2, 2, 3, 5, 4, 3, 0, 1, 1, 2, 2

Draw a tally chart and frequency table to show this distribution. Plot a vertical bar chart. What was the modal score?

Solution

The range of goals is found from the given list i.e. scores of 0 to 5. The list is then checked and each individual score is marked with a small stroke in the tally box opposite the appropriate score (see the table below).

Goals	Tally	Frequency
5	I	1
4	II	2
3	III	3
2	⊞ II	7
1	IIII	4
0	⊞	5
	Total	22

THE FIFTH STROKE IN EVERY BATCH OF FIVE IS USUALLY SHOWN AS A DIAGONAL LINE THROUGH THE OTHER FOUR. THIS MAKES COUNTING THE TALLY STROKES EASY.

The total for each tally box is then entered in the frequency table.

Figure 12.1 shows the bar chart.

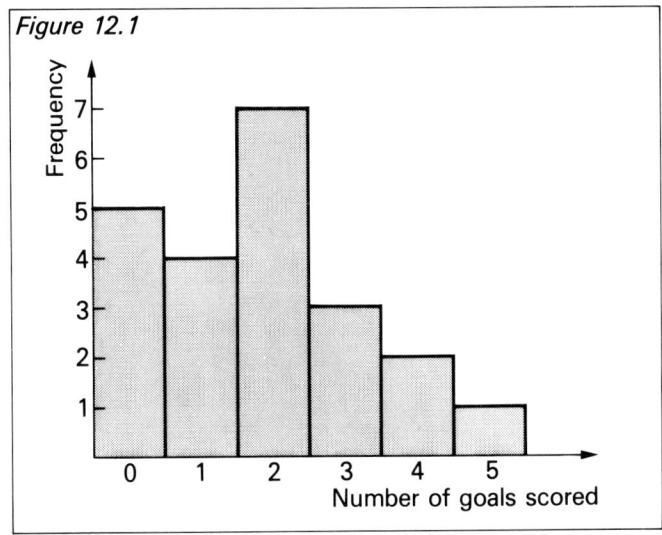

Figure 12.1

The modal score is two goals since this occurs most frequently (7 times).

NOTE THAT THE VERTICAL AXIS IS ALWAYS THE FREQUENCY.

Example 2

A biology experiment consisted of measuring the lengths of all the worms which were extracted from a cubic metre of soil taken from an urban garden. The lengths in centimetres of the fifty worms that were found are recorded as follows.

3.2, 4.5, 6.8, 4.1, 7.5, 8.6, 8.7, 12.5, 5.8, 9.3, 6.2, 5.6, 4.2, 7.5, 7.8, 9.4, 5.2, 10.5, 11.3, 9.7, 8.4, 5.9, 6.8, 7.2, 8.4, 10.1, 15.5, 6.2, 3.9, 5.8, 9.9, 4.7, 6.9, 8.5, 7.6, 10.4, 11.9, 4.4, 7.8, 6.2, 3.7, 6.7, 7.6, 8.3, 4.5, 5.9, 6.8, 6.6, 8.9, 7.5

Draw a frequency table using class intervals of 3.0–3.9, 4.0–4.9, etc. and draw a vertical bar chart to illustrate the information. What is the modal class interval? Estimate the mode using your graph.

Solution

Set out the tally chart and frequency table as shown below using the stated class intervals.

Class interval (length of worm in cm)	Tally	Frequency
3.0–3.9	III	3
4.0–4.9	IIII I	6
5.0–5.9	IIII II	7
6.0–6.9	IIII IIII	9
7.0–7.9	IIII III	8
8.0–8.9	IIII II	7
9.0–9.9	IIII	4
10.0–10.9	III	3
11.0–11.9	II	2
12.0–12.9	I	1
	Total	50

Now plot the vertical bar chart (see Figure 12.2).

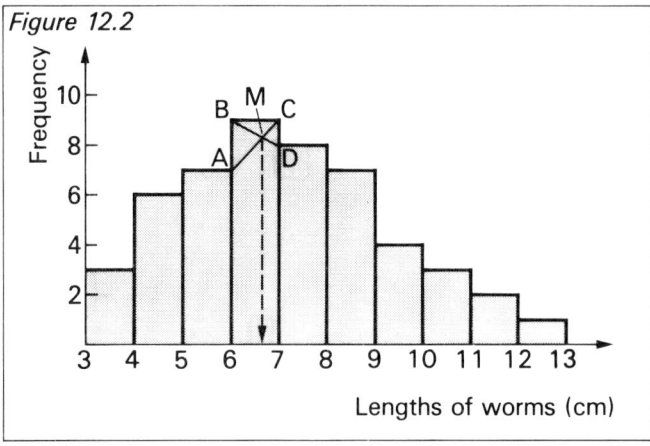

Figure 12.2

The modal class is 6.0–6.9. The mode can be estimated from the modal class. Join A to C. and B to D. Where these lines cross can be called point M. Now drop a vertical line from M to the horizontal axis and read off. The mode can be seen to be approximately 6.7 cm.

THIS METHOD ONLY GIVES A ROUGH ESTIMATE OF THE ACTUAL MODAL AVERAGE.

Exercise 12.1

1 The number of late buses leaving a terminus each day for a month were as follows.

3, 4, 7, 2, 10, 5, 9, 6, 2, 9, 5, 3, 4, 6, 7, 8, 9, 2, 3, 5, 4, 5, 7, 2, 5, 2, 5, 3, 9, 5

Construct a tally chart/frequency table and draw a histogram to display this information. What was the modal number of late buses during the month?

2 A large hospital recorded the following number of 'accidents in the home' for each day during the month of June.

3, 2, 0, 3, 1, 0, 2, 2, 1, 4, 1, 0, 2, 4, 0, 5, 3, 2, 1, 2, 3, 0, 1, 1, 0, 2, 2, 3, 1, 2

Draw a tally chart/frequency table and construct a vertical bar chart. What was the modal number of accidents in June?

3 A restaurant manager recorded the following numbers of complaints made during the first twenty-one days after opening a new steak bar.

0, 1, 3, 0, 2, 1, 1, 2, 0, 1, 4, 1, 2, 1, 1, 2, 0, 1, 2, 1, 1

Draw a tally chart and frequency table for this information. Plot a histogram and use it to find the modal number of complaints.

4 A survey of car ages, which is based on the last letter of the registration plate, was done at a car park containing forty cars. The last letter of each plate was as follows.

P, R, R, X, W, V, V, S, T, L, W, R, W, M, L, M, S, N, X, W, W, P, T, V, M, X, V, L, M, P, K, T, W, V, T, R, N, R, T, W

Draw a tally chart and frequency table for this data. Then draw a histogram and find the modal registration letter.

If the letter K represents 1971, find the year of manufacture of the modal car. (Note: The letters O, Q and U are not used.)

5 A small theatre has one hundred rows and each row has ten seats. The following is a list of thirty consecutive nights' attendances in terms of rows filled.

51.9, 15.7, 25.5, 30, 23.2, 42.5, 56.3, 60.9, 55.5, 58.2, 47.2, 28.5, 39.2, 45.7, 52.1, 67.7, 79.5, 74.5, 72.2, 36.3, 42.9, 51.7, 36.2, 44.9, 57.6, 62.2, 44.7, 18.9, 27.3, 38.7

Draw a tally chart/frequency table using class intervals of 10–19.9, 20–29.9, 30–39.9, etc. Draw a histogram to illustrate this information. What is the modal class interval? Estimate the modal attendance. (Hint: There are ten seats per row.)

6 A post office worker recorded the masses of a batch of thirty-five parcels in kilograms. They were as follows:

2.3, 4.1, 0.7, 2.1, 0.5, 3.3, 2.4, 0.1, 2.7, 3.9, 0.6, 2.7, 2.8, 2.7, 0.9, 0.8, 1.5, 2.1, 1.6, 1.5, 0.2, 3.2, 4.2, 3.7, 2.4, 2.1, 0.7, 2.2, 1.9, 1.2, 0.7, 0.6, 1.9, 1.4, 2.1

Using class intervals of 0–0.5, 0.6–1.0, 1.1–1.5, etc., draw a frequency table and plot a bar chart to display the information. If an automatic sorter removes parcels whose masses are greater than 1.5 kg but less than 4 kg, how many of this batch of parcels will remain? Estimate the modal mass of the parcels.

7 Forty owners of a new model of car were asked to do an average fuel consumption test over a period of one year. The following is the list of observed fuel consumptions in miles per gallon.

27, 22, 23, 26, 27, 26, 25, 21, 29, 21, 25, 30, 26, 30, 26, 25, 18, 24, 22, 26, 23, 22, 29, 22, 25, 26, 31, 22, 19, 24, 26, 27, 29, 33, 21, 27, 23, 26, 24, 25

Draw a tally chart/frequency table using class intervals of 18–20, 21–23, 24–26, etc. Plot the bar chart to display this information. What is the modal class interval? Estimate the modal fuel consumption.

8 The number of live births at a large maternity hospital for each week of a year are recorded as follows.

14, 18, 21, 23, 19, 7, 19, 5, 8, 9, 30, 24, 21, 15, 11, 8, 18, 12, 19, 5, 12, 15, 17, 21, 23, 19, 11, 17, 16, 35, 30, 26, 17, 13, 7, 22, 20, 23, 11, 29, 26, 15, 16, 13, 12, 28, 19, 25, 17, 15, 25, 7

Draw a tally chart/frequency table using class intervals of 1–5, 6–10, 11–15, etc. Draw a histogram and find the modal class interval. Estimate the modal number of births per week.

9 The following is a list of heights in centimetres of 30 three-year-old children attending a playschool. Draw a frequency table using class intervals of 87–92, 93–98, 99–104, etc., and draw a bar chart to illustrate the information. Estimate the modal height of the children.

102, 105, 106, 111, 95, 90, 101, 100, 106, 98, 119, 97, 96, 88, 114, 110, 102, 103, 96, 100, 114, 92, 87, 106, 115, 97, 120, 104, 102, 95

10 The numbers of torn or defaced pages of fifty books in the children's section of a library were as follows.

2, 1, 3, 3, 0, 2, 6, 5, 2, 4, 9, 2, 0, 0, 6, 5, 2, 7, 8, 9, 3, 0, 1, 2, 1, 2, 0, 2, 1, 0, 3, 4, 2, 4, 7, 0, 2, 1, 2, 5, 2, 3, 2, 4, 1, 5, 6, 2, 0, 2

Draw a tally chart/frequency table for this information and plot an appropriate histogram. What was the modal number of faults?

11 The average number of matches in a box was claimed by the manufacturers to be fifty. A group of school children decided to check this claim by counting the number of matches in a sample of forty different boxes. Their findings were recorded as follows.

51, 49, 50, 51, 49, 47, 50, 49, 51, 50, 46, 52, 50, 51, 52, 49, 53, 47, 46, 49, 47, 50, 50, 52, 46, 49, 54, 50, 52, 53, 52, 48, 50, 47, 52, 51, 50, 46, 47, 48

Draw a tally chart/frequency table for the full range of matches recorded and plot a histogram to represent the information. Are the manufacturers correct in their claim? If so, which of the three averages, mode, median or mean did they quote?

12 A train stopped at twenty stations on its journey. The length of time for which the train was stationary at each station was recorded in minutes and seconds. The following is a list of these observations.

5, 5.21, 7.05, 8.46, 9.30, 6, 10.08, 6.51, 6.40, 7, 7.49, 8.21, 9.33, 6, 6.35, 7.11, 8.59, 7.23, 6.07, 9.43

Draw a frequency table to record the information using class intervals 5–5.59, etc. Draw a histogram and use it to answer the following questions.
(a) Between which two times did the train stand in most of the stations? (b) Between which two times did the train stand in the least number of stations?

The cumulative frequency curve (ogive)

THIS TYPE OF CURVE IS USED TO REPRESENT A FREQUENCY DISTRIBUTION AND USUALLY HAS A DISTINCTIVE 'S' SHAPE.

Example 3

A machine that cuts steel bars has a poor tolerance and a 10 cm bar can be cut any length from 9.6 cm to 10.5 cm. The table below shows the length distribution of 100 bars cut by the machine.

Length of bar (cm)	Frequency
9.6	2
9.7	4
9.8	7
9.9	20
10	40
10.1	15
10.2	6
10.3	3
10.4	2
10.5	1

Draw a cumulative frequency curve. What is the median average length of bar? Find the upper and lower quartiles and the semi-interquartile range. What percentage of the bars are less than 9.8 cm? If any bars longer than 10.3 cm are rejected, what percentage is this?

Solution

Draw the cumulative frequency table as shown below.

Length of bar (cm)	Frequency	Cumulative frequency
9.6	2	2
9.7	4	6
9.8	7	13
9.9	20	33
10	40	73
10.1	15	88
10.2	6	94
10.3	3	97
10.4	2	99
10.5	1	100

Plot the cumulative frequency curve using the values from the table. The range is plotted along the horizontal axis and the cumulative frequency along the vertical axis (see Figure 12.3).

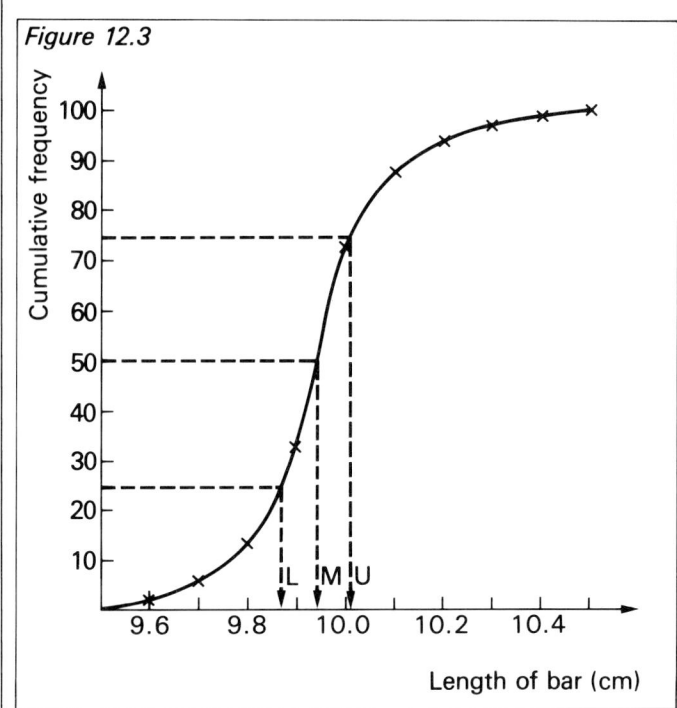

Figure 12.3

THE <u>MEDIAN</u> AVERAGE LENGTH OF BAR CORRESPONDS TO HALF-WAY UP (OR DOWN) THE CUMULATIVE FREQUENCY AXIS (THE 50TH BAR).

Draw a line horizontally through the 50 until it touches the curve. From this point on the curve take a line vertically downwards until it touches the horizontal axis. The median value (*M*) can be seen to be approximately 9.95.

The values of U and L can be found in the same way as the median and are read off Figure 12.3 as

$$U = 10.01$$
$$L = 9.87$$

Semi-interquartile range $= \dfrac{U - L}{2}$

$$= \dfrac{10.01 - 9.87}{2}$$

$$= \dfrac{0.14}{2}$$

$$= 0.07$$

The percentage of bars less than 9.8 cm can be found by first locating 9.8 cm on the horizontal axis. Take a line vertically upwards until it meets the curve. Then take a line horizontally until it cuts the cumulative frequency axis and read this value off. Hence

bars less than 9.8 cm = 13%

The percentage of bars less than 10.3 cm can be found in the same way. So using the graph

bars less than 10.3 cm = 97%

If all bars greater than 10.3 cm are rejected then $(100 - 97)\% = 3\%$ are rejected.

Exercise 12.2

1 The table shows the age distribution of 100 shoppers entering a supermarket.

Age range	Frequency
0–10	8
11–20	12
21–30	15
31–40	20
41–50	18
51–60	12
61–70	10
71–80	5

Draw a cumulative frequency curve and use it to estimate the median average age of the shoppers. What percentage of shoppers were under the age of 65?

2 In an English exam, 100 pupils scored the following percentages.

% score	Frequency
1–10	2
11–20	3
21–30	15
31–40	18
41–50	25
51–60	17
61–70	12
71–80	8

Draw a cumulative frequency curve and use it to estimate the median mark. If the pass mark was 40%, how many failed? What percentage of pupils scored more than 65%?

3 A cost comparison based on the same piece of hi-fi equipment sold in fifty different shops is recorded in the table below.

Cost in £	Frequency
161–165	3
166–170	9
171–175	19
176–180	11
181–185	8

Plot an ogive and use it to estimate the median cost of the hi-fi. What percentage of equipment is over £168?

4 A survey was carried out which was based on the number of people who got on at each bus stop of a long suburban route in the 'rush hour'.

Number of passengers	Frequency
0–2	13
3–5	12
6–8	6
9–11	3
12–14	2

Draw a cumulative frequency curve and find the median number of passengers per stop. Use your graph to estimate the number of stops at which less than seven people got on the bus.

5 Sixty motorbikes took part in time-trials over a two-mile cross-country track. Their times are recorded in the table below.

Time (min : seconds)	Frequency
4 : 01–4 : 30	7
4 : 31–5 : 00	12
5 : 01–5 : 30	17
5 : 31–6 : 00	11
6 : 01–6 : 30	8
6 : 31–7 : 00	5

Draw an ogive and find the median time for the course. If those who took longer than 5 : 15 were eliminated in the first round, how many riders progressed to the next round?

6 The management of a clothing factory employing sixty people is analysing the number of faults per worker in a week's production of garments. Their findings are recorded in the table below.

Number of faults	Frequency
0–3	20
4–6	15
7–9	12
10–12	10
13–15	3

Draw a cumulative frequency curve and use it to estimate the median number of faults per worker. If the company offers an incentive bonus to those who make less than four faults in a week, how many workers are likely to benefit?

7 A survey of fifty houses on a small estate showed the following number of units of electricity consumed in a winter quarter.

250, 245, 248, 230, 269, 255, 270, 220, 231, 244, 244, 230, 235, 220, 290, 230, 233, 250, 247, 262, 241, 275, 260, 280, 210, 309, 240, 230, 258, 263, 288, 253, 271, 308, 244, 300, 229, 272, 294, 288, 263, 230, 219, 260, 229, 227, 256, 232, 257, 292

Draw a tally chart and cumulative frequency table using class intervals of 210–219, 220–229, etc. Draw an ogive and find the median usage of electricity. Find the upper and lower quartiles and the semi-interquartile range. What percentage of consumers used more than 275 units of electricity?

8 The infant mortality rate per 1000 live births for all the countries of Europe is recorded as follows.

14, 12, 10, 11, 17, 11, 8, 16, 21, 16, 12, 20, 15, 11, 23, 21, 16, 33, 26, 35, 44, 24, 21, 19, 38, 14, 30, 28

Draw a tally chart and a cumulative frequency table using class intervals of 6–10, 11–15, etc. Draw an ogive and use it to estimate the median infant mortality rate. Estimate the upper and lower quartiles and the semi-interquartile range. How many countries have an infant mortality rate less than 23?

Basic probability

The probability that an event will occur is governed by the number of successful outcomes and the total number of possible outcomes. Provided no two outcomes occur at the same time (called mutually exclusive outcomes) then the following formula holds.

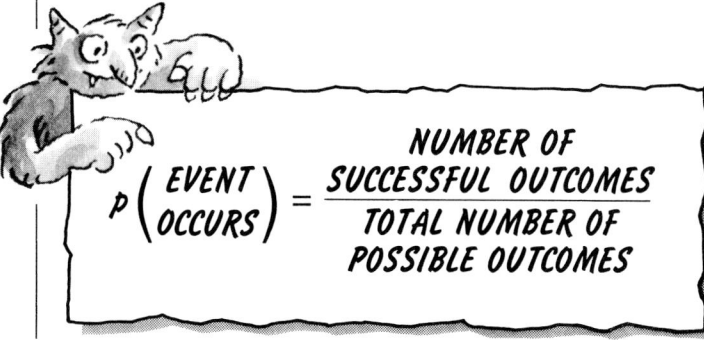

$$P\left(\begin{array}{c}\text{EVENT}\\\text{OCCURS}\end{array}\right) = \frac{\text{NUMBER OF SUCCESSFUL OUTCOMES}}{\text{TOTAL NUMBER OF POSSIBLE OUTCOMES}}$$

All probabilities are expressed as fractions which lie between the limits of 0 and 1.
An event having a probability of 0 can never occur (impossibility).
An event having a probability of 1 is certain to occur (certainty).

Example 4

A normal die is numbered one to six and each number is equally likely to turn up. Find the probability of obtaining the following on a single throw of the die:
(a) a two   (b) an even number   (c) a number less than five.

Solution

(a) $\text{p}(2) = \dfrac{\text{number of times a two occurs on the die}}{\text{total number of values on the die}}$

$\text{p}(2) = \dfrac{1}{6}$

(b) p(even number)

$= \dfrac{\text{number of even numbers on the die}}{\text{total number of values on the die}}$

$= \dfrac{3}{6}$ (2, 4, 6 are the even numbers)

$= \dfrac{1}{2}$

(c) p(number less than five)

$= \dfrac{\text{total of all numbers less than five}}{\text{total number of values on the die}}$

$= \dfrac{4}{6}$ (1, 2, 3, 4 are numbers less than 5)

$= \dfrac{2}{3}$

Example 5

Tim's freezer has nine orange and seven chocolate lollies. Each one is equally likely to be chosen. If one lolly is picked at random find the probability that it is
(a) orange   (b) chocolate.

Assuming that a chocolate lolly is chosen and eaten, what is the probability that a second lolly picked at random will be   (c) orange   (d) chocolate?

Solution

(a) $\text{p(orange lolly)} = \dfrac{\text{number of orange lollies}}{\text{total number of lollies}}$

$= \dfrac{9}{16}$

(b) $\text{p(chocolate lolly)} = \dfrac{\text{number of chocolate lollies}}{\text{total number of lollies}}$

$= \dfrac{7}{16}$

The sum of these two probabilities is 1 since it is a certainty that either an orange or a chocolate lolly would be picked.

If one chocolate lolly is eaten then the problem becomes one of choosing from nine orange and only six chocolate lollies. The total number of lollies is now only fifteen.

(c) $\text{p(orange lolly)} = \dfrac{\text{number of orange lollies}}{\text{total number of lollies}}$

$= \dfrac{9}{15}$

$= \dfrac{3}{5}$

(d) $\text{p(chocolate lolly)} = \dfrac{\text{number of chocolate lollies}}{\text{total number of lollies}}$

$= \dfrac{6}{15}$

$= \dfrac{2}{5}$

Exercise 12.3

1 In a four-horse race, the probabilities that Cruiser, Pussyfoot and His Lordship will win are $\frac{1}{5}$, $\frac{1}{4}$ and $\frac{3}{10}$ respectively. What is the probability that the fourth horse, Spiritfingers, will win? Which horse would start the favourite? (Assume that one of the four horses must win.)

2 A machine can only fail to operate if its G-pin or its K-pin fractures. If a G-pin fractures twice as often as a K-pin, what is the probability that when the machine fails next it will be the K-pin that has fractured?

3 A card is selected randomly from a normal pack of well shuffled playing cards (no jokers). Find the probability that the card drawn is   (a) black   (b) a queen   (c) a red seven   (d) the ace of clubs   (e) a heart   (f) a picture card   (g) a numbered card which is divisible by 5.

4 A book contains 30 pages (numbered 1–30). What is the probability that any page chosen at random will contain a 5 digit in it?

5 A small staff car park contains 5 Ford, 8 British Leyland, 3 Vauxhall and 4 foreign cars. Find the probability that any car chosen at random will be (a) Vauxhall (b) British Leyland.

6 A box of shoe rejects at a factory has 14 men's and 21 women's shoes in it. What is the probability that the first shoe chosen at random will be a man's?

7 What is the probability that a bingo number (1–90) drawn at random has a 4 digit in it?

8 A box contains x red beads, $2x$ white beads and $3x$ green beads. What is the probability that a bead chosen at random is (a) green (b) white?

9 A football side has a probability of losing of $\frac{2}{3}$ and a probability of drawing of $\frac{1}{4}$. What is the probability that they might win?

10 A pack of bulbs contains 15 hyacinth, 20 daffodil and 25 crocus. What is the probability that a bulb chosen at random is (a) daffodil (b) crocus?

11 Janet has four times as much chance of passing maths as Dougie. What is the probability that Dougie will pass if Janet has a $\frac{4}{5}$ chance of passing?

12 A car-hire firm buys British and foreign cars in the ratio 3 : 2. What is the probability that a car offered at random to a customer will be British?

13 A tube containing twelve sweets is made up from the following colours: 3 red, 4 green and 5 orange. If the sweets are packed randomly what is the probability that the first one chosen is (a) orange (b) red?

Assuming that the first sweet taken is red and that it is eaten, what is the probability that the second sweet chosen is (c) orange (d) red?

14 Examine the word STATISTICS. If one of the letters is chosen at random, what is the probability that it is (a) T (b) I?

15 A school committee consists of 2 first-year pupils, 3 second-year pupils, 4 each from the third, fourth and fifth years and 5 from the sixth form. One person is chosen at random to be the chairman. What is the probability that he or she will be (a) a first-year pupil (b) a fifth-year pupil (c) a sixth former?

16 A delivery of sausages is as follows: 45 kg pork, 30 kg beef and 25 kg tomato. What is the probability that one kilogram pack chosen at random will be (a) beef (b) pork (c) tomato?

17 A bookcase for returned books at a library has authors with the first letter of their surnames arranged as follows.

K, F, M, F, A, C, S, K, C, F, B, T, J, S, W, W, L, D, F, R

If one of these books were chosen at random what would be the probability that the author's surname started with (a) R (b) S (c) F?

18 A page of a child's book was analysed to find the number of letters in each word. The information was as follows

number of letters 1 2 3 4 5 6 7
number of words 6 5 10 8 6 4 1

What is the probability that a word chosen at random has (a) 6 letters (b) 3 letters (c) 4 letters?

19 A box of fireworks contains 6 roman candles, 5 catherine wheels, 7 volcanoes, 1 atomic whizzbang, 3 golden fountains and 2 snowstorms.
(a) A firework is chosen at random. What is the probability that it is (i) a roman candle (ii) an atomic whizzbang (iii) a snowstorm?
(b) If an atomic whizzbang and a volcano are set off, what is the probability that the next firework chosen at random will be (i) a roman candle (ii) a volcano (iii) an atomic whizzbang?

20 A toffee manufacturer mixes his flavours chocolate, cream and mint in the ratio 3 : 5 : 7. What is the probability that a toffee chosen at random will be (a) chocolate (b) cream?

Combined probabilities

(a) Two events are said to be mutually exclusive if they cannot occur at the same time.

FOR TWO MUTUALLY EXCLUSIVE EVENTS A AND B, THE PROBABILITY THAT EITHER EVENT A OR B WILL OCCUR IS FOUND BY ADDING THEIR RESPECTIVE PROBABILITIES TOGETHER.

THUS $p(A \text{ OR } B) = p(A) + p(B)$

(b) Two events are said to be independent of one another if there is no way that the outcome of one event can affect the outcome of the other.

FOR TWO _INDEPENDENT_ EVENTS A AND B, THE PROBABILITY THAT EVENT A _AND_ EVENT B WILL OCCUR IS FOUND BY MULTIPLYING THEIR RESPECTIVE PROBABILITIES TOGETHER.

THUS $p(A \text{ AND } B) = p(A) \times p(B)$

Example 6

The probability that Hotshot Harris will score a goal is $\frac{3}{8}$. The probability that his striking partner Bomber Brown will score is $\frac{1}{3}$. What is the probability that in the next match (a) either will score (b) both will score (c) neither will score?

Solution

(a) p(Harris or Brown scoring) = p(Harris) + p(Brown)

$$= \frac{3}{8} + \frac{1}{3}$$

$$= \frac{9+8}{24}$$

$$= \frac{17}{24}$$

(b) p(Harris and Brown scoring) = p(Harris) × p(Brown)

$$= \frac{3}{8} \times \frac{1}{3}$$

$$= \frac{1}{8}$$

(c) p(neither will score) = p(Harris will not score) × p(Brown will not score)

$$= \frac{5}{8} \times \frac{2}{3}$$

$$= \frac{10}{24}$$

$$= \frac{5}{12}$$

Example 7

A five-sided spinner numbered 1 to 5 is spun at the same time as a fair die numbered 1 to 6 is thrown. Their total score is then noted. Draw a _probability space diagram_ to show all possible outcomes and use it to find the following probabilities: (a) total score is 7 (b) the score on the die and spinner is the same (c) the total is less than 5 (d) the score on the die and spinner is the same _and_ the total score is less than 5.

Solution

The probability space diagram is shown in Figure 12.4.

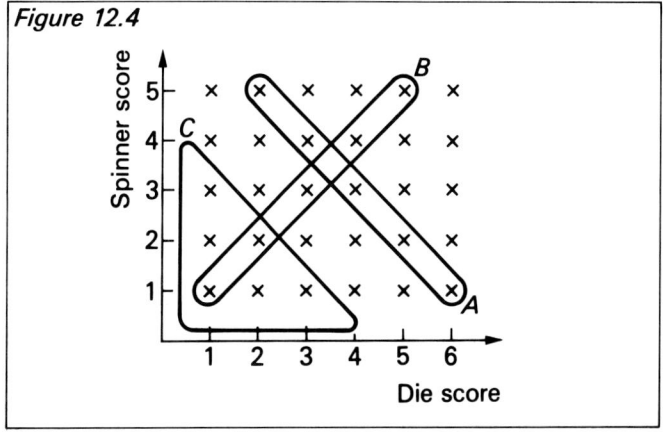

Figure 12.4

Set A is 'total score of 7' (5 outcomes).
Set B is 'score on the die and spinner is the same' (5 outcomes).
Set C is 'total less than 5' (6 outcomes).

(a) p(total score of 7) = $\frac{5}{30} = \frac{1}{6}$

(b) p(same score on spinner and die) = $\frac{5}{30} = \frac{1}{6}$

(c) p(total less than 5) = $\frac{6}{30} = \frac{1}{5}$

(d) p(same score on spinner and die and total score less than 5) = $\frac{2}{30} = \frac{1}{15}$

Number of successful outcomes = $n(B \cap C) = 2$

Example 8

A man's gas and electricity meters are always read on the same day each quarter. However the probability that the gas bill will arrive first is $\frac{3}{5}$ and the electricity bill is $\frac{2}{5}$.

Draw a tree diagram to show all the possible outcomes of the arrival of the bills for the first three quarters of the year. Use the tree diagram to find the following probabilities: (a) the gas bill arrives first each quarter (b) the gas bill arrives first only once (c) the gas bill arrives first at least twice.

Solution

The tree diagram is shown in Figure 12.5.

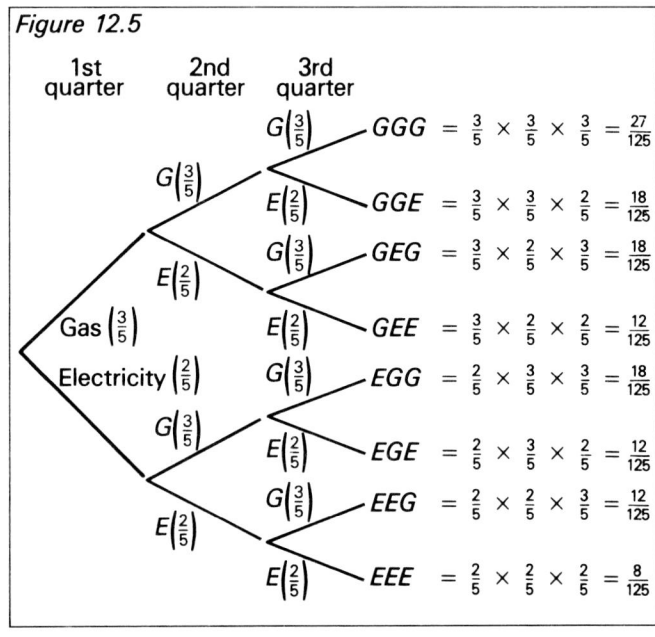

Figure 12.5

(a) p(gas bill arrives first each quarter)
$$= p(G) \text{ and } p(G) \text{ and } p(G)$$
$$= p(G) \times p(G) \times p(G)$$
$$= \frac{3}{5} \times \frac{3}{5} \times \frac{3}{5}$$
$$= \frac{27}{125}$$

(b) p(gas bill arrives first only once)
$$= p(GEE) \text{ or } p(EGE) \text{ or } p(EEG)$$
$$= \frac{12}{125} + \frac{12}{125} + \frac{12}{125}$$
$$= \frac{36}{125}$$

(c) p(gas bill arrives first at least twice)
$$= p(GGE) \text{ or } p(GEG) \text{ or } p(EGG) \text{ or } p(GGG)$$
$$= \frac{18}{125} + \frac{18}{125} + \frac{18}{125} + \frac{27}{125}$$
$$= \frac{81}{125}$$

Exercise 12.4

1 A card is drawn from a well shuffled pack (no jokers) and its value is noted. The card is then replaced. A second card is then drawn and its value noted. Find the probability that on the two successive draws from the pack the following are obtained (a) two picture cards (b) an ace first and a king second (c) a heart first and a black card second (d) a red king and an eight in any order (e) a number card less than four followed by a red picture card. (Assume an ace counts as a one.)

2 Three 10-pence coins are tossed together and their outcomes (heads or tails) are noted. Construct a simple table to show all possible outcomes. Find the probability that the following will occur (a) three heads (b) two heads and a tail (c) at least two tails.

3 A bag contains eight cubes, three are black and five are white. One cube is drawn at random. Its colour is noted and it is replaced. A second cube is then drawn. Find the probability that the following will occur (a) they are two blacks (b) their colours are different (c) their colours are the same. (A tree diagram may help.)

4 A small car park has two Ford cars, three British Leyland and one Vauxhall. All cars are equally likely to leave first. Two cars leave and do not return. Find the probability that the following has occurred (a) both were British Leyland (b) a Ford left first and a Vauxhall second (c) both cars were the same make.

5 A canteen offers three main dishes: a salad, a roast and fish. The probability that Sam will choose a salad is $\frac{1}{4}$, a roast is $\frac{5}{12}$ and fish is $\frac{1}{3}$. Draw a tree diagram to show all possible outcomes for two days choice of main courses. Use the tree diagram to find the probability that the following will occur. Sam chooses (a) fish twice (b) no salads (c) salad at least once (d) the same dish twice.

6 Two five-sided spinners (numbered 1 to 5) are spun together. Draw a probability space diagram to show all possible outcomes. Find the probability that the following will occur (a) the total is seven (b) the total is less than six (c) the score is a double.

7 A fruit machine has two reels. Each reel has ten symbols, two orange, three lemons and five pears. (a) What is the probability that the following will turn up on one reel only (i) an orange (ii) a lemon (iii) a pear?

(b) Complete the tree diagram shown in Figure 12.6 and use it to find the following probabilities
(i) orange and lemon in any order (ii) exactly two lemons (iii) at least one pear (iv) the two fruits are the same.

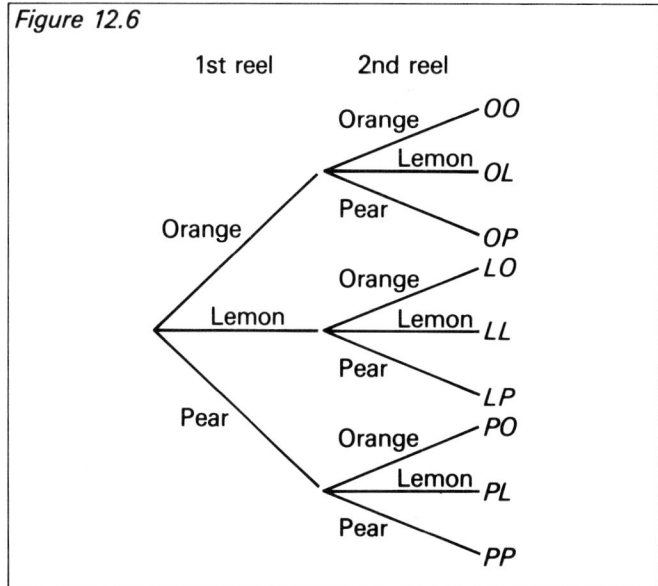

Figure 12.6

1st reel   2nd reel

Orange
 Orange — OO
 Lemon — OL
 Pear — OP

Lemon
 Orange — LO
 Lemon — LL
 Pear — LP

Pear
 Orange — PO
 Lemon — PL
 Pear — PP

8 The probability that Dad will be given socks, a tie or a handkerchief for Christmas are $\frac{2}{15}$, $\frac{1}{5}$ and $\frac{2}{3}$ respectively. What is the probability that Dad will receive (a) a tie or a handkerchief (b) a pair of socks and a tie (c) a pair of socks or a tie or a handkerchief (d) a pair of socks and a tie and a handkerchief?

9 It is only possible for four boys to win a prize. The probabilities are as follows: Richard $\frac{1}{8}$, Tom $\frac{1}{3}$, Mike $\frac{1}{6}$, Matthew x. Which boy is most likely to win a prize? What is the probability that (a) both Mike and Tom win a prize (b) Richard or Matthew wins a prize?

10 In a fall during a climbing accident the probability that a man will break an arm, a leg and a rib are $\frac{1}{20}$, $\frac{1}{4}$ and $\frac{1}{5}$ respectively. What is the probability of breaking (a) an arm and a leg (b) a rib or an arm (c) an arm and a leg and a rib (d) none of these?

11 In a single day's hunt to catch wild animals for a zoo, the probability of catching an elephant, a rhinoceros and a zebra are given as $\frac{1}{30}$, $\frac{1}{6}$ and $\frac{1}{3}$ respectively. What is the probability of catching in one day (a) an elephant and a zebra

(b) a rhinoceros or an elephant (c) a zebra and an elephant and a rhinoceros (d) none of these?

12 The probabilities that two brothers Jock and Jake will pass an exam are $\frac{2}{5}$ and $\frac{1}{3}$ respectively. Find the probabilities that (a) Jock and Jake pass
(b) Jock or Jake passes (c) they both fail
(d) only Jock passes.

13 Two eight-sided spinners (numbered 1 to 8) are spun together. Draw a probability space diagram to show all possible outcomes. Use the diagram to find the following probabilities. (a) The score is a double. (b) The total is greater than ten.
(c) The total is eight. (d) The total is a prime number.

14 All the picture cards (12) are taken from a pack and shuffled carefully. A card is then drawn from this small pack of picture cards and not replaced. A second card is also drawn. Find the probability that the following will occur. (a) Two queens are drawn. (b) Two spades are drawn. (c) A red card is followed by a black. (d) Both cards are the same suit.

15 A bag contains ten sweets, six chocolates and four toffees. A sweet is drawn at random from the bag and eaten. A second and third sweet are also chosen and eaten. Complete the tree diagram in Figure 12.7 and use it to answer the following questions. What is the probability that the following sweets are chosen (a) three toffees (b) exactly two chocolates (c) at least one chocolate
(d) three sweets of the same flavour?

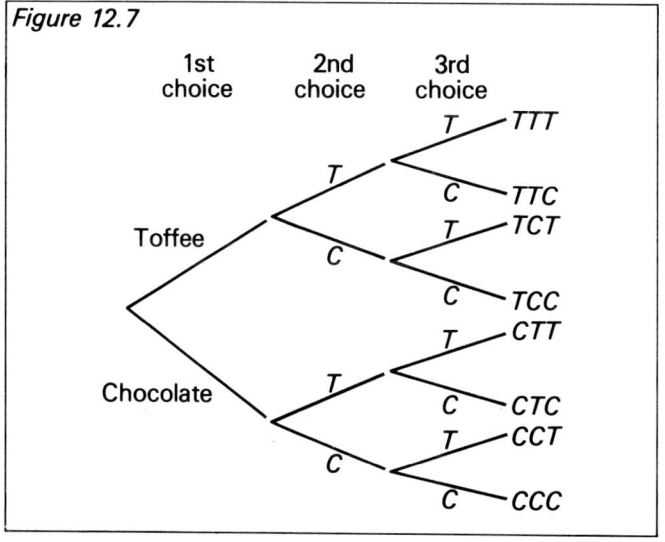

Figure 12.7

1st choice   2nd choice   3rd choice

Toffee
 T
  T — TTT
  C — TTC
 C
  T — TCT
  C — TCC

Chocolate
 T
  T — CTT
  C — CTC
 C
  T — CCT
  C — CCC